PHENOMENOLOGICAL PSYCHOLOGY

PHAENOMENOLOGICA
COLLECTION FONDÉE PAR H.L. VAN BREDA ET PUBLIÉE
SOUS LE PATRONAGE DES CENTRES D'ARCHIVES-HUSSERL

103

PHENOMENOLOGICAL PSYCHOLOGY

edited by

JOSEPH J. KOCKELMANS

PHENOMENOLOGICAL PSYCHOLOGY

The Dutch School

edited by

JOSEPH J. KOCKELMANS

1987 **MARTINUS NIJHOFF PUBLISHERS**
a member of the KLUWER ACADEMIC PUBLISHERS GROUP
DORDRECHT / BOSTON / LANCASTER

Distributors

for the United States and Canada: Kluwer Academic Publishers, P.O. Box 358, Accord Station, Hingham, MA 02018-0358, USA
for the UK and Ireland: Kluwer Academic Publishers, MTP Press Limited, Falcon House, Queen Square, Lancaster LA1 1RN, UK
for all other countries: Kluwer Academic Publishers Group, Distribution Center, P.O. Box 322, 3300 AH Dordrecht, The Netherlands

Library of Congress Cataloging in Publication Data

Phenomenological psychology.

(Phaenomenologica ; 103)
Bibliography: p.
Includes index.
1. Phenomenological psychology. 2. Phenomenological psychology--Netherlands. I. Kockelmans, Joseph J., 1923- . II. Series. [DNLM: 1. Psychology. BF 204.5 P541]
BF204.5.P48 1987 150.19'2 87-5718

ISBN 90-247-3501-7
ISBN 90-247-2339-6 (series)

PRINTED IN THE NETHERLANDS

Table of Contents

Preface

Over the past decades many books and essays have been written on phenomenological psychology. Some of these publications are historical in character and were designed to give the reader an idea of the origin, meaning, and function of phenomenological psychology and its most important trends. Others are theoretical in nature and were written to give the reader an insight into the ways in which various authors conceive of phenomenological psychology and how they attempt to justify their views in light of the philosophical assumptions underlying their conceptions. Finally, there are a great number of publications in which the authors do not *talk about* phenomenological psychology, but rather try *to do* what was described as possible and necessary in the first two kinds of publications. Some of these attempts to do the latter have been quite successful; in other cases the results have been disappointing.[1] This anthology contains a number of essays which I have brought together for the explicit purpose of introducing the reader to the Dutch school in phenomenological psychology.

The Dutch school occupies an important place in the phenomenological movement as a whole. Buytendijk was one of the first Dutch scholars to contribute to the field, and for several decades he remained the central figure of the school. Van Lennep, van den Berg, Strasser, and Linschoten are other outstanding members. Although Plessner was of German descent and a philosopher by profession, I have included an essay by him also, in view of the fact that he taught at the University of Groningen and in his own way has made an important contribution to the Dutch school in phenomenological psychology.[2]

I have tried in this anthology to bring together a number of essays which actually show the phenomenological-psychological method at work. Some selections included here have been taken from works already available in English, whereas others were expressly translated for this volume. In the translating of these selections I have attempted to combine two principles: the translations were to stay as close as possible to the originals; and the translations were to be rendered in "good English." I hope the reader finds these two principles combined successfully here.

1. Cf. Herbert Spiegelberg, *Phenomenology in Psychology and Psychiatry*. (Evanston: Northwestern University Press, 1972). A selected bibliography on the subject can be found on pp. 369-400.
2. *Ibid.*, p. xxxvi, 287-289.

As for the questions involving what phenomenological psychology is, what it attempts to accomplish, the methods to be used, how these methods are to be employed, how this kind of psychology is to be related to empirical research within the realm of psychology, how phenomenological psychology is to be related to philosophy, etc., I must refer the reader to the substantial literature available in these areas. Yet in order to give the reader a first orientation in regard to the relevant issues, I have included two essays which deal with these questions. And to facilitate further study I have added a selected bibliography in which are included only those books and articles which in my opinion offer an objective, sound, and understandable introduction to the subject. I have listed, also, a few publications which describe the origin and the development of phenomenological psychology.

Some of the essays contained in this book have been taken from the vast realm of hermeneutico-phenomenological psychology. That is to say, not all essays found here try to give a "mere description" of psychic phenomena; some of the essays, in dealing with psychic phenomena, are the results of attempts to bring to light and articulate "that which shows itself in the very way in which it shows itself from itself" *interpretatively*. Hermeneutic phenomenology was first suggested by Heidegger; his conception of phenomenology has since been adopted by many phenomenologists, philosophers as well as psychologists.

The choice I have made, however, by no means implies that "pure eidetic descriptions" are psychologically irrelevant, nor that they are unimportant. The abundance of available literature necessitated a selective choice. In view of the fact that one can easily find eidetic descriptions in the works of Husserl, Scheler, Pfänder, Edith Stein, Aron Gurwitsch, Sartre, Merleau-Ponty, and many others, I decided to emphasize in this anthology contributions from the area of hermeneutico-phenomenological psychology.

The hermeneutic psychologist attempts to show "the things themselves," but in so doing he always presupposes a realm of meaning, a world, within which these things show themselves the way they do. The authors represented in this book have taken our "Western world" as the relevant world here. In showing "the things themselves" each author has his particulr stance in this world. The authors do not claim that they have shown "the things themselves" from all possible points of view; they *interpret* the meaning the relevant things have in our world from their own point of view in that world; they claim merely that anyone else who is willing to adopt this point of view will confront the same phenomena and will confront them in the same way they have. That such an approach to psychic phenomena is by no means 'subjective' is something of which the reader can easily be convinced by reading the selections which follow. Obviously in some instances the reader will disagree with an author's interpretation. In these cases the authors invite the reader to engage in a dialogue with them for the purpose of eliminating mistakes and inadequacies, and in this way to come to a genuine, intersubjectively acceptable understanding of the relevant phenomena.

The authors included in this book frequently make use of poetry and literature. The reason is threefold: 1) The authors believe that many great poets and novelists have, indeed, seen something very important and have spoken of it in a remarkably adequate way. 2) Sometimes novels and poems illustrate a point on which the phenomenologist wishes to focus attention. And 3) most importantly, often an appeal to poetry and literature is almost unavoidable in that poetic language with its use of symbolism is able to refer beyond the realm of what can be said "clearly and distinctly." In other words, most of the authors represented here feel that in human reality there are certain phenomena which reach so deeply into a man's life and the world in which he lives that poetic language is the only adequate way through which *to point* to and so to make present a meaning which we are unable to express clearly in any other way. However, it is important to note that no author included in this volume has used literary works as a substitute for the work he himself has tried to accomplish. That is, poems and novels do not "prove" anything. But both can be enormously helpful in bringing certain phenomena closer to us and thus in making us "understand" them, helping us to understand ourselves and the world in which we live.

Finally, I wish to call the reader's attention to another point of major significance. It is often said that phenomenological and hermeneutical psychology should eliminate all forms of empirical psychology. That this notion is a flagrant misinterpretation of the genuine intention of these authors can be shown easily. First of all, no *leading* phenomenologist has ever made this claim. On the contrary, all of them have argued explicitly that what we call "psychology" is a complex of various disciplines, each with its own typical methods: empirical psychology uses empirical methods, eidetic phenomenology employs descriptive methods, and hermeneutic phenomenology uses interpretative methods. Thus in the view of the leading phenomenologists, empirical psychology is possible and necessary and no phenomenological or hermeneutic psychology can be substituted for it. A psychologist must know "the facts" just as he must understand "their meaning" in our Western world. This is why he must learn to work as an empirical psychologist and to think about the meaning of what his research reveals to him. Husserl, Heidegger, Jaspers, Sartre, and many other phenomenologists distinguish explicitly between empirical and phenomenological psychology and go to great lengths to explain how these different psychological disciplines are to be related to one another. And their ideas have been shared by the phenomenologists represented in this book. The reader should know, also, that many of the authors represented here have made important contributions to empirical as well as to phenomenological psychology. Buytendijk and Linschoten in particular have shown that and how the various approaches can go hand in hand, complementing each other.

Grateful acknowledgment is made to the publishers who have granted permission to reprint selections from copyrighted material. I should like to thank my wife who has assisted me in choosing the selections and in solving some of the difficult problems connected with the translations.

Joseph J. Kockelmans
The Pennsylvania State University

PART I

What is Phenomenological Psychology

1

Husserl's Original View on Phenomenological Psychology*

JOSEPH J. KOCKELMANS

Some forty years ago Edmund Husserl spoke publicly for the first time of a new phenomenological psychology. He saw this new psychology as a discipline that was destined to play an important role in the already established empirical psychology as well as in philosophy. Subsequently under the influence of his ideas an extensive phenomenological psychological movement began to spread through various European countries. When a careful analysis of this phenomenological movement is made, one becomes aware of a number of clearly distinguishable currents and schools, all of which claim Husserl as their origin. The truth is, however, that only a very few psychologists actually use Husserl's concepts without making major modifications. Furthermore, many psychologists talk about phenomenology without stipulating precisely what is meant by the term. To compound the difficulties there is noticeable in phenomenological literature a frequent failure to make a clear distinction between Husserl's thought and that of other phenomenologists such as Scheler, Heidegger, Jaspers, Merleau-Ponty, Sartre and Binswanger.

The confusion becomes understandable when one realizes that in psychology as well as in sociology and anthropology different trends of thought did in fact develop in the absence of clearly formulated philosophical premises. In addition, it has been virtually impossible for most writers to see phenomenological problems in historical perspective. Also, and most importantly, until very recently Husserl's thought has been available only to a privileged few who had access to his manuscripts. Knowledge about his ideas as communicated by these secondary sources has shown considerable divergence on many points. All of these factors constitute difficulties when one tries to determine just what is to be understood by phenomenological psychology. It is apparent that a reasonable solution cannot be found by using some kind of largest common denominator of all the existing phenomenologies as a point of departure for a description of phenomenological psychology. A meaningful formulation can be reached only through a careful historical investigation into the essential differences underlying the different phenomenologies.

It is, in fact, now possible to make a fresh start with important historical and critical investigations, for thanks to the indefatigable efforts of the Husserl-Archives we have access to the complete text of Husserl's last work *Krisis* and to all his other important contributions to phenomenological psychology from 1920 to 1938.[1]

However, the specific task with which we are presently concerned is not that of an historical and critical appraisal of Husserl's own development as a phenomenological psychologist between 1891 and 1938, for anyone who is interested in this issue may find several such studies at his disposal.[2] It is against the background of these historical studies, however, that certain questions shall be dealt with briefly here: How did Husserl arrive at his views regarding phenomenological psychology? What task did he envision for this new discipline? What are its essential aspects? How does this psychology relate to empirical psychology on one hand and to philosophy on the other? Only after these questions have been answered will it be possible to evaluate Husserl's work in the light of the recent developments that have been proposed by various protagonists of phenomenological psychology. This approach, I believe, will clear up an embarrassing situation by disclosing the real perspectives that Husserl's thought has opened up for present-day psychology.

In order to explain the problems mentioned as clearly as possible I shall focus attention primarily on Husserl's final view as explained in *Phenomenological Psychology* and *Krisis*. However, a few remarks on Husserl's earlier view seem to be a necessary introduction to his final standpoint.

1. THE GENESIS OF THE IDEA OF A PHENOMENOLOGICAL PSYCHOLOGY

Psychologism and Husserl's Criticism

It was in Leipzig in 1876 that Husserl first became acquainted with the psychology of his day. He had started his scientific career with a thorough study of mathematics and physics, but after only a few years of work he had come into contact with psychology under the influence of Wundt, Paulsen, Brentano, and Stumpf. Now Husserl's interest in philosophy, at first minimal, was eventually aroused by Brentano; and thus, under the influence of his psychological studies with Brentano and Stumpf and also through his contact with the works of Stuart Mill, Spencer, Locke,

1. Edmund Husserl, *Die Krisis der europäischen Wissenschaften und die transzendentale Phänomenologie. Eine Einleitung in die phänomenologische Philosophie*, herausgegeben von Walter Biemel (Husserliana, Band IV) (The Hague: Nijhoff, 1954). English translation by David Carr: *The Crisis of European Sciences and Transcendental Phenomenology* (Evanston: Northwestern University Press, 1970). For the foregoing remarks see: Stephan Strasser, *Phenomenology and the Human Sciences: A Contribution to a New Scientific Ideal* (Pittsburgh: Duquesne University Press, 1963), pp. 245-248.

2. Hermann Drüe, *Edmund Husserls System der phänomenologischen Psychologie* (Berlin: Walter de Gruyter, 1963); Joseph J. Kockelmans, *Edmund Husserl's Phenomenological Psychology: A Historico-critical Study* (Pittsburgh: Duquesne University Press, 1967); Aron Gurwitsch, *Studies in Phenomenology and Psychology* (Evanston: Northwestern University Press, 1966).

and Hume, he very soon became entangled in empiricism and psychologism. He would later have to shake off psychologism's tendency to found the objects of mathematics, logic, epistemology, theory of value, and so on, on subjective psychical experiences.

Initially Husserl's special interest in philosophy was focused on the philosophical foundation of mathematics and logic. In his first book, *Philosophy of Arithmetic* (1891),[3] he tried to derive the fundamental concepts of arithmetic from certain psychological acts. It was in this book that Husserl defended a kind of psychologism by advancing the thesis that the epistemological foundations of arithmetic must be given ultimately by empirical psychology. Influenced by Frege's critique of this book and by a publication of Natorp, Husserl very soon came to the conclusion that psychologism is inadmissible. In 1895 in his lectures at Halle, Husserl began to present his own critique of psychologism which was later published in 1900 in the first volume of the *Logical Investigations.*[4] In the second part of this book Husserl explained that, although the ideal objects studied by mathematics and logic have a being of their own, there must be a typical correlation between these ideal objects, belonging to the realm of logic and pure mathematics, and our psychical, lived experiences as the activities which constitute them. Although the fundamental ideas of a phenomenological philosophy are already implicit in the second volume of the *Logical Investigations*, it was not until 1907, in a series of lectures entitled *The Idea of Phenomenology,*[5] that Husserl was able to formulate these basic insights in a systematic way.

From 1907 on we find psychology as a constant pole of comparison in Husserl's explanations of the meaning of his phenomenological philosophy. Practically speaking, in every work dealing with the foundations of phenomenological philosophy, Husserl tries to explain his view on empirical psychology and to describe the difference and the relationship between these two sciences.

Phenomenological Psychology

Between 1911 and 1913 the problem concerning the relation between empirical psychology and phenomenological philosophy had Husserl's special attention again. Gradually it became clear to him that it is possible and even necessary to bridge the gap between empirical psychology and transcendental phenomenology with the help of a completely new science which was called "rational psychology", or

3. Edmund Husserl, *Philosophie der Arithmetik. Psychologische und logische Untersuchungen*, Erster Band (Halle a.S.: Pfeiffer, 1891).
4. *Logische Untersuchungen*, 2 vols. (Halle a.S.: Niemeyer, 1900-1901). English translation of the second edition by J.N. Findlay: *Logical Investigations*, 2 vols. (London: Routledge and Kegan Paul, 1970).
5. Edmund Husserl, *Die Idee der Phänomenologie. Fünf Vorlesungen* (1907), herausgegeben von Walter Biemel (Husserliana, Band II) (The Hague: Martinus Nijhoff, 1950). English translation by William P. Alston and George Nakhnikian: *The Idea of Phenomenology* (The Hague: Martinus Nijhoff, 1950).

"eidetic psychology" first, and "phenomenological psychology" afterward. This view is explicitly expressed for the first time in *Ideas* (1913),[6] although the first traces of these insights are already found in his *Logos* article.[7]

It seems that between 1913 and 1923 Husserl was especially concerned with the so-called constitutional problems. As a result of these investigations it became apparent to him that his explicitation of the meaning of phenomenological psychology as presented in *Ideas* was not adequate, in that the distinction between phenomenological psychology and transcendental philosophy had not yet been founded in a radical way. It became apparent also that the new science, phenomenological psychology, was of such importance for the development of the empirical psychology of his time that a radical investigation of it seemed to be demanded. It was for this reason that he dealt with the topic in different lecture courses between 1923 and 1927. The results of these investigations were published posthumously in 1963 in a book entitled *Phenomenological Psychology.*[8] In this work phenomenological psychology is described as an aprioristic, eidetic, intuitive, purely descriptive, and intentional science of the psychical, which remains entirely within the realm of the natural attitude.[9] Here Husserl tries to found the necessity of such a new kind of psychology by pointing to the fact that traditional empirical psychology still lacks a systematic framework of basic concepts grounded in the intuitive clarification of the psychical essences. Whatever psychology has accumulated and is still accumulating by way of measuring and experimentation concerning objective correlations is wasted as long as there is no clear grasp of what it is that is being measured and correlated. According to Husserl, phenomenological psychology is destined to supply the essential insights needed to give meaning and direction to the research presented under the title "empirical psychology."

Exactly the same ideas are found in Husserl's *Enclyclopædia Britannica* article,[10] his *Amsterdam Lectures,*[11] and the *Cartesian Meditations.*[12] It is in these publica-

6. Edmund Husserl, *Ideen zu einer reinen Phänomenologie und phänomenologischen Philosophie.* Erstes Buch: *Allgemeine Einführung in die reine Phänomenologie*, herausgegeben von Walter Biemel (Husserliana, Band III) (The Hague: Martinus Nijhoff, 1950). English translation by F. Kersten: *Ideas Pertaining to a Pure Phenomenology and to a Phenomenological Philosophy* (The Hague: Martinus Nijhoff, 1982).
7. Edmund Husserl, "Philosophie als strenge Wissenschaft," in *Logos*, 1 (1910-1911), pp. 289-341. English translation by Quentin Lauer, *Edmund Husserl: Phenomenology and the Crisis of Philosophy* (New York: Harper & Row, 1965), pp. 69-147. See also: *Cross Currents*, 6(1956), pp. 228-246 and pp. 324-44.
8. Edmund Husserl, *Phänomenologische Psychologie* (1925), herausgegeben von Walter Biemel (Husserliana, Band IX) (The Hague: Martinus Nijhoff, 1962). English translation by John Scanlon: *Phenomenological Psychology*, Summer Semester, 1925 (The Hague: Martinus Nijhoff, 1977).
9. *Ibid.*, pp. 46-51.
10. "Phenomenology," in *Encyclopædia Britannica*, 14th ed. (London, 1927), vol. 17, cls. 699 702.
11. "Amsterdamer Vorträge," in *Phänomenologische Psychologie*, pp. 302-349.
12. Edmund Husserl, *Cartesianische Meditationen und Pariser Vorträge*, herausgegeben von Stephan Strasser (Husserliana, Band I) (The Hague: Martinus Nijhoff, 1950). English translation by Dorion Cairns: *Cartesian Meditations. An Introduction to Phenomenology* (The Hague: Martinus Nijhoff, 1960).

tions that Husserl mentions for the first time that a special reduction is essential and indispensable for a phenomenological psychology, too. In his last book, *Krisis*, Husserl returns to this point and in a detailed and minute inquiry tries to determine the very nature of this reduction and to found its necessity.[13] In these investigations he comes to very strange conclusions regarding the relation between psychology and phenomenology.[14]

2. HUSSERL'S VIEW ON EMPIRICAL PSYCHOLOGY. "EIDETIC PSYCHOLOGY"

Spiegelberg has rightly pointed to the fact that Husserl never opposed psychology as a whole, but only certain types of psychology which he indicated with the epithets "naturalistic" and "objectivistic." With these expressions Husserl refers to psychologies which, in mistaken imitation of the physical sciences, tried to get rid of the essential features of psychological phenomena.

To understand Husserl's point of view in regard to empirical psychology, we must realize that the psychology of his day consisted of a combination of psychophysical and physiological investigations carried out to determine quantitatively and experimentally the relationships between objective stimuli and subjective responses. Brentano and James had evidently proposed several very important remarks to correct this fundamental misconception of psychology, but neither had materially altered the classical conception. The same holds true for Gestalt-psychology, which was also still trapped by the prejudices of objectivism and scientism. According to Husserl, Dilthey was the first scientist who clearly saw the fundamental mistakes in the leading psychological schools: naturalism and objectivism. But even Dilthey was not able to indicate a new and correct way to psychology.[15]

Husserl explained this view on empirical psychology for the first time in his article *Philosophy as a Strict Science*. Since it provided the basis for the preliminary description of the new phenomenological psychology that Husserl introduced in his *Ideas*, a short summary of the most important insights proposed in the *Logos* article is in order.

First Misconception of Traditional Psychology: No Pure Analysis

In the first part of his *Logos* article Husserl attempted to explain the necessity of a phenomenological philosophy. There he described his phenomenology as a science of consciousness, but distinguished it from psychology as a *natural* science about

13. Edmund Husserl, *Krisis*, pp. 238-269.
14. Joseph J. Kockelmans, *Edmund Husserl's Phenomenological Psychology*, Chapter VI and the literature quoted there.
15. Edmund Husserl, *Phänomenologische Psychologie*, pp. 4-13. See also: Herbert Spiegelberg, *The Phenomenological Movement: A Historical Introduction*, 2 vols. (The Hague: Martinus Nijhoff, 1960), Vol. I, pp. 150-151.

consciousness. Ultimately he concluded that there is an evident strong relationship between phenomenology and psychology since both are concerned with consciousness, though in a different way and according to different orientations. Psychology is concerned with empirical consciousness, that is, with consciousness as an empirical being in the real world, whereas phenomenology is concerned with pure consciousness.[16]

From this it will be clear that, in principle, psychology is more closely related to philosophy than the other sciences, which do not deal with consciousness at all. However, what has been said here of a necessary relationship between psychology and philosophy does not apply to modern empirical psychology. For the fundamental conviction of this psychology is that pure analysis and description of the data which immediately manifest themselves in immanent intuition are to be put aside in favor of certain indirect psychologically relevant facts brought to light by observation and experiment. Such a psychology does not see that without an essential analysis of conscious life these facts are deprived of their real meaning. In other words, although it is true that empirical psychology is able to bring to light valuable psychophysical facts and norms, it nevertheless remains deprived of a deeper understanding and a definitive scientific evaluation of these facts so long as it is not founded in a systematic science of conscious life which investigates the psychical as such with the help of "immanent" intuitive reflection. By the very fact, therefore, that experimental psychology considers itself as already methodologically perfect, it is actually unscientific wherever it wishes to penetrate to a *real* psychological understanding. On the other hand, it is equally unscientific in all those cases where the lack of clarified concepts of the psychical as such leads to an obscure formulation of problems and consequently to merely apparent solutions. The experimental method is indispensable, particularly where there is a question of fixing intersubjective connections of facts. But this does not alter the fact that it presupposes what no experiment can accomplish, namely, the analysis of conscious life itself.[17]

Some psychologists, such as Stumpf and Lipps, had recognized this defect of empirical psychology and, in the manner of Brentano, had tried to undertake thorough analytical-descriptive investigations of psychical experiences. The results of these investigations were denied recognition by most of the experimental psychologists, who disdainfully called them "scholastic analyses." The only reason for this depreciation, however, was that Brentano, Stumpf, and Lipps took ordinary language as the starting point of their investigations. But if one reads these investigations it becomes clear immediately that Brentano, Stumpf, and Lipps do not derive any judgment at all from word-concepts, but rather penetrate to the phenomena themselves which immediately present themselves to man's intuitive reflection.

Be this as it may, it is evident that the fixation of scientific language presupposes a complete analysis of the original phenomena, and that as long as that has not been

16. Edmund Husserl, *Philosophie als strenge Wissenschaft*, pp. 299-322 (90-110).
17. *Ibid.*, pp. 302-323 (92-94).

accomplished the progress of the investigations remains to a great extent in the realm of vagueness and ambiguity.[18]

Second Misconception of Traditional Psychology: Imitation of Physics

In the reactions against this so-called Scholasticism it is very often brought out that "empty" word-analyses are meaningless and that one has to question the things themselves and to go back to experience, which alone can give sense and meaning to our words. The question is, however, what is to be understood here by "the things themselves," and what kind of experience is it to which we have to return in psychology. Are they perhaps the answers the psychologist gets from his "clients" or subjects of experimentation? Or is his interpretation of their answers the experience we must look for? Every experimental psychologist will say, evidently, that the primary experience lies in the subjects and that an interpretation of this experience presupposes certain self-perceptions of the psychologist which – whatever they may be – in any case are not "introspections."[19]

Despite some exaggeration, there is something in this view which is unquestionably right. But there is also a fundamental error in this psychology, for it puts analyses realized in empathetic understanding of others' experiences, and analyses based on one's own formerly unnoticed experiences, on the same level with the analysis characteristic of natural science, in the belief that it is an experimental science of the psychical in fundamentally the same way as natural science is the experimental science of the physical. In so doing, however, it overlooks the specific character of consciousness and the psychical data.

Most psychologists believe they owe all their psychological knowledge to experience. Nevertheless the description of the naïve empirical data, along with an immanent analysis which goes hand in hand with this description, is effected with the help of psychological concepts whose scientific value will be decisive for all further methodological steps. These concepts, however, remain by the very nature of the experimental method constantly untouched, but nevertheless enter into the final empirical judgments which claim to be scientific. On the other hand, the value of these concepts was not present from the beginning, nor can it originate from the experience of the subjects or of the psychologists themselves. Logically it can be obtained even from no empirical determinations whatsoever. And here is the place for phenomenological, eidetic analyses.[20]

What has been constantly muddled in empirical psychology since its beginnings in the eighteenth century is the deceptive idea of a scientific method modeled after that of the physico-chemical sciences. The British "associationists" as well as the German "experimentalists" were convinced implicitly that the method of all empirical sciences, considered in its universal principles, had to be one and the same; therefore, that it ought to be the same in psychology as in the natural sciences. Just

18. *Ibid.*, pp. 303-305 (95-96).
19. *Ibid.*, pp. 305-309 (96-98).
20. *Ibid.*, pp. 309-312 (98-102).

as metaphysics suffered for a long time from an imitation of the geometrical and physical methods (in the work of Descartes and many others), so psychology in the same way has suffered from an unacceptable simulation of the physical sciences. It is not without significance that the fathers of experimental psychology (Fechner, von Helmholtz, and Wundt) were physiologists and physicists. Be this as it may, it is clear that in following these lines the typical characteristics of the psychical phenomena must be denied. The true method has to follow the nature of the things to be investigated, not our prejudices and preconceptions.[21]

Since all psychological knowledge presupposes essential knowledge of the psychical, and since such knowledge cannot be obtained by means of physical procedures, it is evident that only phenomenological analyses can give us a correct solution for the problems mentioned. It is the fundamental error of modern psychology that it has not recognized the necessity of a phenomenological method. For only a really radical and systematic phenomenology, carried out not incidentally and in isolated reflections, but in exclusive dedication to the extremely complex and confused problems of consciousness and executed in an attitude free from all naturalistic prejudices, can give us a real understanding of the psychical. Only then will the plenitude of empirical facts and the interesting laws which have been gathered bear their real fruit as the result of a critical evaluation and psychological interpretation. Then, too, will it become clear in what sense psychology stands in close relationship to philosophy.[22]

3. PHENOMENOLOGICAL PSYCHOLOGY: ITS RELATION TO EMPIRICAL PSYCHOLOGY AND PHENOMENOLOGICAL PHILOSOPHY

Introduction

We have seen in the preceding discussion that according to Husserl we shall only be able to correct the fundamental mistakes of traditional empirical psychology by means of a phenomenological psychology which will fill the gap between philosophy and empirical psychology. Hence we must try now to describe the nature of this phenomenological psychology as accurately as possible, and to indicate the relations between this phenomenological psychology and empirical psychology, as well as its relation to phenomenological philosophy.

We have seen also that in the gradual process of attempting to determine the nature of phenomenological psychology as it is related to empirical psychology and transcendental phenomenology, Husserl's thinking was in a constant state of evolution toward what may be called its final phase, which (with the probable exception of his description of the phenomenological-psychological reduction) was reached about 1928. For this reason a short survey of the most significant ideas in Husserl's

21. *Ibid.*, pp. 312-314 (102-103).
22. *Ibid.*, pp. 314-322 (103-110).

Amsterdam Lectures, which contained for the first time the explanation of this final view, will be presented now.[23]

At the start of these lectures Husserl states that around 1900, as philosophy and psychology struggled for a strictly scienctific method, a new science was born, and, at the same time, a completely new method for philosophical and psychological inquiry came into being. This science was called phenomenology because its new method had its origin in a radicalization of the phenomenological method, the need for which had long been felt, and which had even been applied in some instances by physicists and psychologists such as Mach, Hering, Stumpf, and Brentano. The radicalization of these methodological conceptions, however, led not only to a completely new method in psychology and epistemology, but simultaneously to a new way of handling typically philosophical, foundational problems, so that a new kind of scientific pursuit emerged.

In the further development of this new phenomenology it became necessary to make a fundamental distinction between phenomenological psychology as the foundational science for all psychological disciplines, and transcendental phenomenology, which in the realm of philosophy was designed to take over the function of first philosophy and to subject the foundations of philosophy itself to a radical investigation.[24]

Phenomenological Psychology: Its Subject Matter, Method, and Function

Present-day psychology is the science of the real psychological events which occur in the concrete domain of the spatio-temporal world. In this context "psychical" is to be understood as the ego and all that necessarily is connected with this ego, namely, all ego-centered experiences such as perceiving, thinking, willing, etc. The "psychical" manifests itself in immediate experience as a non-self-sufficient realm of being because it appears only in man and animals, which according to another more fundamental sphere of being are also physical realities. This is why psychology may be considered a branch of the more concrete anthropology and zoology which deal also with the physical and psychophysical aspects of these living beings.[25]

It is an essential characteristic of the world of experience taken as a whole that it differentiate itself in an open infinity of concrete individual realities. Furthermore, it is also an essential trait of every individual reality that it, irrespective of all possible superstructures, possess bodily-physical reality. Accordingly man always has the possibility of abstracting from everything that is non-physical in order to consider each real being and also the world as a whole merely as "physical nature." Already in the domain of the merely physical there is a determinate, essential, structural regularity and legality of the world of experience. Not only does each

23. Edmund Husserl, "Amsterdamer Vorträge," in *Phänomenologische Psychologie*, pp. 302-349.
24. *Ibid.*, pp. 302-303.
25. *Ibid.*, p. 303.

concrete, intramundane being possess its own nature, its bodily-physical reality, but all mundane bodies whatsoever co-constitute an harmonious unity governed by the universal, unifying form of spatio-temporality. By means of a consistently abstracting experience one can focus attention only and exclusively on the physical in intramundane beings and the world of experience as a whole. Afterward it is possible to build upon this experience a theoretical science which is complete in itself, namely, natural physical science, at once comprising chemistry, physical zoology, and biology.[26]

The question now is to what extent (in another one-sided orientation toward the psychical which as self-sufficient never occurs in the same world), a continuously consistent experience and a theoretical inquiry founded on it are possible which, proceeding from the psychical to the psychical, never take the physical as such into consideration. In other words, in addition to the pure empirical natural sciences, in how far is a pure psychology possible? Without further investigation it is clear that factually existing psychology as an empirical science of facts never will be able to become a pure science of merely psychical facts free from any physical datum. For, no matter how far the pure psychological experience and the theory founded on it could reach, it is certain that the merely psychical about which each intends to speak possesses its spatio-temporal determinations in the real world, and in its real factualness is determinable only by means of spatio-temporal determinations. Spatio-temporality, however, originally and essentially belongs to nature taken as physical nature. Everything non-physical, such as the psychical, possesses a spatio-temporal position only because of its foundation in the physical "bodiliness." Thus it becomes clear that it will be forever impossible within the realm of empirical psychology to delineate theoretically a pure psychological investigation from a psychophysical inquiry. In other words, within the domain of empirical psychology as an objective science of facts, it is absolutely impossible to constitute a pure empirical science of the mere psychical as a separate and independent discipline with a clear-cut task, because in empirical psychology which has to deal with concrete real beings, a reference to the psychophysical and the physical seems to be essential and necessary.[27]

However, a pure psychological inquiry is not completely impossible; and for an empirical psychology which strives for a really and strictly scientific character, such a psychology is in fact indispensable. For without a pure psychology it is impossible to obtain the strictly scientific concepts in which the very essence of the psychic phenomena can be expressed truthfully. And since to these concepts belong also those which delineate the universal and necessary essential form of the psychical as such, the necessity of an aprioric science of the essence of the psychical as such makes itself felt. Such a science is not parallel to physics, but to a pure natural science which focuses attention on the necessary a priori of any imaginable "nature" as such, and in part is already materialized in a pure science of time, pure geometry, and pure mechanics.[28]

26. *Ibid.*, pp. 303-304.
27. *Ibid.*, pp. 304-305.
28. *Ibid.*, p. 305.

It is not easy to bring aprioric truths to the fore. As really essential truths they originate in apodictical insight only from the initial source of experience. How can these truths be faithfully brought to light and uncovered? They can be fruitfully considered only in a methodical study and by an accurate unveiling of their all-sided horizons. For only in this way is it possible to uncover the psychical as such in a truly original and concrete manner. In such a procedure, in which the psychical is held to manifest itself in its proper and essential selfhood, that which is concretely experienced functions only as an example. One has to focus attention here first and foremost on that which in the changing and varying examples manifests itself as invariable and abiding.

Above all, therefore, the exemplary, possible, and actual experiences are of importance; and for a scientific investigation of these experiences a determinate method is indispensable. We have to bear in mind here, however, that the psychical cannot be discovered in any experience except by reflection or "perversion" of the natural attitude. We are accustomed to concentrating upon the things, thoughts, and values of the moment, and not upon the psychical acts of experience and the psychical as such. The psychic life and all our psychical acts of experience, with all their different modalities, properties, and horizons, are revealed only by reflection; and such a reflection can be practiced on every concrete, actual, or possible experience. Furthermore we have to realize that everything which comes to light in reflection possesses the typical characteristic of being intentional. Intentionality manifests itself as the essential trait of psychic life in the strict sense of the term. Whenever we speak of appearances, we are referred to subjects to which something appears and to certain aspects of their psychic life according to which an appearance as an appearance-of-something occurs and, therefore, we are also and even necessarily referred to what appears in and through those appearances.[29]

In a certain sense we could say that in every psychic experience something appears to the subject in question insofar as this subject is conscious of it. From this perspective the phenomenality, as the proper trait of the appearances and of that which appears, manifests itself then as the fundamental characteristic of the psychical as such, so that from the same perspective pure psychology can be called "phenomenology," or even better, aprioric phenomenology. Such a psychology evidently must deal also with the ego-subjects in their individual and social aspects, but only insofar as they are subjects of such a phenomenality.[30]

Phenomenologic-Psychological Reduction

But let us return now to our original problem: How can a pure phenomenological experience be brought into light truthfully? Here phenomenological experience is to be understood as only that reflection in which the psychical as such becomes accessible in the way briefly indicated above. In this context we must first ask how this experience can be performed methodically in such a way that by means of its

29. *Ibid.*, p. 305.
30. *Ibid.*, pp. 307-308.

purity it brings to the fore what is essentially characteristic of the psychical as such. The term "purity" has a double meaning here. In the first place, this experience must be *pure* in the sense of free from all the psychophysical and the physical with which it is essentially connected. All that a physical and psychophysical experience could teach us must here be thematically left out of consideration, so that we have to restrict ourselves to a pure phenomenological experience in order to try to explicitate only what in this experience is given immediately and as such. There are evidently many problems here. For how is a pure phenomenological experience to be performed, and how can one proceed from such an experience to a uniform and purely psychical field of experience which even *ad infinitum* would not imply anything non-essential to the psychical as such? Moreover, this experience must be "pure" in a second sense in that it must be free from all prejudices which spring from other scientific spheres of experience and which could blind us to that which phenomenological reflection immediately offers us.[31]

The difficulties necessarily connected with these two requirements are so great that to this date an intentional analysis has never been performed properly in modern psychology. Even Brentano did not succeed in this task although he gave a central place to the concept of intentionality in psychology and tried to lay the foundations of psychology in a systematic and descriptive inquiry of consciousness.[32]

Traditional psychology's confused attempt to imitate the methods of the natural sciences[33] derives from the equalization of immanent temporality and objective real time. Objective time is the "extensional" form of all objective realities and, in particular, of their structural foundation, namely physical nature. The psychical experiences, taken individually and also in their totality, do not in themselves possess the unitary form of coexistence and succession which are characteristic of the spatio-temporal as such. Characteristic of the psychical experiences is the form of "flowing"; according to their very essence they are flowing in the unity of the stream of consciousness as a whole; and this is evidently not a parallel form of spatio-temporality. The intentional analysis of immanent temporality brings to the fore the strong as well as the weak side of the metaphor of the stream, and furthermore makes it perfectly clear that every real analogy between the analysis of consciousness and natural, physical, chemical, and biological analyses and, at the same time, every analogy between consciousness and nature are to be abandoned. "Natural-logical" concepts such as thing, property, whole, part, compound, separation, cause, effect, action, etc., are essentially founded in reality, nature, the *res extensa*; in the realm of the psychical they lose their proper meaning. Taken as psychological terms, they are empty husks or shells because there are left over only formal-logical concepts such as object, quality, etc.[34]

However, there are other fundamental difficulties connected with the realization

31.*Ibid.*, pp. 308-309.
32.*Ibid.*, p. 309.
33.*Ibid.*, pp. 309-310.
34.*Ibid.*, pp. 310-311.

of a consequent and pure phenomenological experience. First of all we have to abandon all the prejudices of tradition as well as the most general self-evidences of logic which are already interpreted too much from the viewpoint of the physical. As a result we have to restrict ourselves to that which phenomenological reflection brings to light as consciousness and the conscious, and to that which manifests itself there in real self-evidence. So we must exclusively adhere in this situation to the phenomenological experience without paying any attention to the factual data which actually appear in it. This experience is first of all self-experience; only in this experience are consciousness and the conscious-ego given in completely original selfhood. All other forms of experience of the psychical are founded in the immediacy of this self-experience; this holds good also for the pure experience of things, the others, and society. Thus from its first conception Husserl explained the method of pure self-experience as the method of a consistently performed phenomenological self-disclosure.[35]

Somehow we must omit from consideration what refers to external experience, which is the original source of every physical consideration. This is difficult to realize, particularly since the psychical belonging to the others is to be excluded also. For every experience is an experience of something, of the world, and every reflection presupposes an immediate experience of something other.[36]

If I as a phenomenologist am to realize a pure phenomenological experience in order to make consciousness, my own conscious life as regards its pure and proper essence, into a universal and consequent theme of investigation and, therefore, first into a domain of pure phenomenological experience, then I have to eliminate as non-psychic-being the real world as a whole (the being-value of which was already accepted a priori in my natural life) from the theme of investigation.[37]

On the other hand, it is evident that all consciousness is indeed consciousness of something, and that in direct experience what we are immediately conscious of is the natural world, the real spatio-temporal world. If that is true, then it is possible to describe a perception and an act of memory according to their very essence without mentioning the fact that they are perception and memory of this determinate object. The consequence is that reflection on consciousness as such does not yet bring to light the psychical in its own pure essence. We must, besides, refrain from our natural belief in the reality of the world. As phenomenologists we must as it were be the "disinterested onlookers" of our own conscious life which only in this way can become a theme of investigation given in experience. Instead of living "in" our own consciousness and being interested in the world given in it, we have only to look upon it just as consciousness of this or that and just as it is thus interested in itself; otherwise the nonpsychic world rather than our own consciousness would be the object of our description. However, within the sphere of such an *epochē* consciousness remains as always consciousness of something, of this or that objectivity. That the conscious datum itself as such comes to the fore in every consciousness of something is essential for consciousness as psychic datum. Within the realm of the

35. *Ibid.*, pp. 311-312.
36. *Ibid.*, p. 312.
37. *Ibid.*, p. 312.

epochē, however, this conscious datum is taken only as such; that which is experienced in every conscious intentional act is – within the realm of the *epochē* – not a being in the real world; this being is taken here only as that toward which consciousness' intention is directed; as reality in the world this object is put between brackets. In this way the phenomenological reduction is, essentially speaking, delineated. It is, however, important to call attention again to the fact that in this reduction not only is the noetical preserved, but also the noematical as an endlessly fruitful theme of phenomenological description. It is precisely through phenomenological reduction that for the first time intentional objectivities can be delivered as such, that is as essential constituents of intentional experiences.[38]

Phenomenological reduction also influences our attitude with respect to the consciousness-ego, because here, too, every *real* animal and *real* human aspect is put between brackets. Just as nature is reduced by means of the reduction to a noematic phenomenon only, so also is the real human ego in the natural attitude reduced to pure psychic life. My "being-a-man" in the real world and my mundane life is maintained only as "meant," that is, as that toward which the intentional conscious acts of intentional life, reduced to the pure psychical, are and continue to be oriented.[39]

The consistent disclosure of the noema can shift toward a consideration and analysis of the correlative noeses. But in addition to these conscious intentional acts the ego-center as such manifests itself as something on and in itself, as the ego of every *cogito*, as an ego which in all these acts is and continues to be phenomenologically identical, as the center of radiation from which all the various and specific ego-acts beam forth. The ego thus manifests itself here as the center from which all acts emanate and toward which all affects flow back. But in both these respects the phenomenological ego-center is an important and extensive phenomenological theme closely connected with every other phenomenological topic.

It is in these analyses concerning the pure life of the ego that the fundamental and essential distinction between the mode of being of consciousness in its phenomenological purity and nature as it is given in our natural attitude clearly manifests itself; this distinction is clearest in the ideality according to which the noematic contents are included in every conscious act. Therefore one can also say that this difference consists in the typical synthesis which makes each consciousness-of into a unity and connects it with other conscious acts, making a unity of *one* consciousness. All forms of synthesis ultimately go back to identity syntheses.[40] Let me try to explain this briefly.

In every conscious act we are directed toward an object, we "intend" it; and reflection reveals this to be an immanent process characteristic of all experiences. To be conscious of something is not an empty having of such a something in consciousness. Each phenomenon has its own intentional structure, which analysis shows to be an ever-widening system of intentionally related, individual com-

38. *Ibid.*, pp. 312-314.
39. *Ibid.*, pp. 314-315.
40. *Ibid.*, p. 316.

ponents. The perception of a house, for instance, reveals a multiple and synthesized intention: a continuous variety in the appearances of the house, according to differences in the points of view from which it is seen and corresponding differences in perspective, and all the differences between the front side actually seen at the moment and the back side which is not seen and which remains, therefore, relatively indeterminate, and yet is supposed to be equally existent. Observation of the stream of these noemata and of the manner of their synthesis, shows that every phase is already in itself a consciousness-of-something, yet in such a way that with the constant entry of new phases the total consciousness, at any moment, is a consciousness of one and the same house. In this it is implied that in every conscious act we are referred to an indeterminate number of similar experiences of the same house and, therefore, that in the noema of a certain individual act there are already implied references to other aspects of the same house which in this individual act are already predelineated as real or possible aspects of it. The same holds true for every conscious intentional act. Here the real essence of an intentional relation becomes manifest: that of which I am conscious in every intentional experience is a noematic pole which refers to an open infinity of always new intentional experiences in which this house would appear as identically the same. This means that the noematic pole is not really but only ideally contained in the different possible experiences.[41]

Phenomenological Psychology

In conclusion we may say that a systematic construction of a phenomenological psychology requires: 1) the description of all the characteristics belonging to the essence of an intentional lived experience and of the most general law of synthesis in particular; 2) the explanation of the characteristic features and forms of the different types of lived experiences which necessarily are found in every consciousness, and all their typical syntheses; 3) the explanation and essential description of the very essence of the universal stream of consciousness; 4) an inquiry into the ego as center of the lived experiences and as pole of all actualities and potentialities of consciousness. When this static description is drawn we must try to analyze and describe the genesis of the life of the personal ego with its universally eidetic laws; thus we must combine a genetic phenomenology with the static type previously described. Our genetic phenomenology must explain the different modes of active and passive genesis, and in regard to the latter, especially the phenomenological, new concept of association. The static and the genetic phenomenology of reason is a special, coherent field of inquiry of a higher level which is probably most important within the realm of transcendental phenomenology.

Finally we must remark again that the validity of all these investigations will obviously extend beyond the particularity of the psychologist's own consciousness. For psychical life may be revealed to us not only in self-consciousness but equally in our consciousness of other selves, and this latter source of experience offers us

41. *Ibid.*, p. 316.

more than a reduplication of what we find in our self-consciousness, for it establishes the difference between "our own" and "the other" which we experience, and presents us also with the characteristics of the "social life." Hence the further task becomes a matter of psychology's revealing the intentions of which the social life consists.[42]

Phenomenological and Empirical Psychology

Let us suppose now that via the phenomenological reduction mentioned above we have put ourselves in the sphere of the pure psychological, and that with the help of intentional analyses and the method of free variation we have gained an insight into the essence of the psychical in its diverse modalities. The aprioric concepts which in this phere are formed through eidetic reduction must then express an essentially necessary style to which every imaginable, factual, and real psychic life is tied. All empirical psychological concepts are governed by these aprioric concepts as well as by their logical "forms," just as analogically such is the case with physics and the general aprioric science of nature. It is self-evident, therefore, that the aprioric truths founded in these aprioric concepts possess an unconditional, normative validity in regard to the regions of being in question, and in this particular case, in regard to the empirical domain of the pure psychical.[43]

In comparing phenomenology with the much more embracing empirical psychology, it must be said that phenomenological psychology is the absolutely necessary fundament for the performance of an exact empirical psychology, a longstanding goal heretofore pursued in attempts to follow the example of the exact physical, natural sciences. It is now necessary that empirical psychology conform to the exactness required by *modern* natural science. Natural science, which was once also a vague, inductive, empirical science, owes its modern character to the a priori system of forms characteristic of nature as such; this system is constituted by pure geometry, pure mechanics, and the pure science of time. By theoretically referring the factual in experience to the a priori of these forms, the originally vague experience is able to participate in the essential necessity, which is the last root of the exactness of the physical sciences.[44]

The methods of natural science and psychology are admittedly quite different, but the latter, like the former, can only reach exactness by means of a rationalization of the essential. This means that in empirical psychology the exactness must be founded in the very essence of the psychical as such. As we have seen, the essence of the psychical as such must be brought to light through the investigations of phenomenological psychology, so that phenomenological psychology has to provide us with the fundamental concepts which, describing the a priori structure of the psychical as such, must govern every possible psychological description.[45]

42. *Ibid.*, pp. 315-321.
43. *Ibid.*, pp. 321-324.
44. *Ibid.*, pp. 324-325.
45. *Ibid.*, p. 325.

Here, however, a typical problem manifests itself. For the a priori of empirical psychology is more extensive than that which is explained by phenomenological psychology. Empirical psychology as a science of the psychical which in the given world manifests itself as a real moment and thus belongs to nature as psychophysical datum is, therefore, also co-founded by the a priori of physical nature. The necessary consequence is that empirical psychology also is based on the empirical and aprioric sciences of nature. Ultimately it is even founded in its own a priori which belongs to the psychophysical as such. In other words, the a priori of empirical psychology is not exclusively phenomenological, for it depends not only on the essence of the psychical but also upon the essence of the physical, and more particularly upon the essence of the psychophysical of organic nature.[46]

Phenomenological Psychology and Transcendental Philosophy

Finally we must try to describe the relation between phenomenological psychology and transcendental phenomenology. On the one hand, psychology, both as an eidetic and empirical discipline, is a "positive" science effected within the realm of the "natural attitude" and, therefore, accepting the world as the ground of all its statements, whereas transcendental phenomenology is completely "unworldly." On the other hand, however, there is a close relationship between phenomenological psychology and transcendental phenomenology in that phenomenological psychology requires only a more stringent re-employment of the formal mechanisms of reduction and analysis to disclose the transcendental phenomena which form the subject matter of transcendental phenomenology. In order to explain this double relationship between phenomenological psychology and transcendental phenomenology we must take our starting point in Husserl's view of the meaning of the transcendental reduction.[47]

Man normally lives in the natural attitude in accordance with which the world around him is at all times known and accepted as a *real fact-world* that has its real being independent of man's consciousness. Phenomenological philosophy invites every philosopher to alter this standpoint radically. This change of attitude is to be performed by the "transcendental reduction." Although this alteration of attitude has a certain similarity with Descartes' methodical doubt, the transcendental reduction is itself completely different from such a doubt. For the attempt to doubt any object in regard to its actually being there necessarily demands a certain suspension of the general thesis of the natural attitude, while the transcendental reduction is not a transformation of the thesis of the natural attitude into its antithesis, nor a transformation of one's certain conviction into a presumption or a real doubt. We do not rule out the thesis; we take it out of the picture only in that we "bracket" it; it itself remains, however, like the bracketed in the brackets. In short, the general thesis of the natural attitude is still experienced as lived, but we choose to make no use of it within the realm of transcendental phenomenology.

46. *Ibid.*, pp. 326-328.
47. *Ibid.*, p. 328.

It can be seen from this description that the transcendental reduction is responsible for putting the general thesis of the natural attitude out of action; whatever is included in this general thesis is placed in brackets with due respect for its nature as "being real." In so doing the real world is not denied; neither is it doubted; the transcendental reduction is used only in the sense of completely barring one from using any judgment that concerns the real spatio-temporal existence of the world out there. That is why all sciences which relate to this natural world are also disconnected in the transcendental reduction, although there is no intention here of objecting to them ultimately. No use is made of their methods and conclusions.

Such a procedure is required for philosophy because the transcendental reduction is the absolutely necessary condition for finding a realm of being that is *apodictically evident*. For what is left over after the reduction is pure consciousness with its pure correlates and its pure ego, insofar as they are not affected by the transcendental reduction. Transcendental reduction, therefore, is the necessary operation which renders pure consciousness and subsequently the whole phenomenological region accessible. It opens up the absolute, apodictically evident region of being – the region of absolute consciousness, of transcendental subjectivity, in which the totality of being, the whole actual and possible universe is contained. After the reduction there is neither a world that *really* exists for man, nor a science which he can accept. In this new attitude the whole world is for man only something *that claims being*. From this point on, the world as a whole and every intramundane being is for each one only a phenomenon instead of something that really exists out there.

By the transcendental reduction we gain possession of the whole stream of our conscious life as it is given in *apodictical evidence*, and of everything meant in its lived experiences, as meant in them purely: the universe of pure phenomena. The being of the pure ego and its pure cogitations, as a being that is necessarily prior in itself, is antecedent to the natural being of the world. The natural, real world is a realm of being whose existential status is secondary; it continually presupposes the realm of "transcendental" being: pure consciousness and its pure cogitata.[48]

Comparing this brief description of the transcendental reduction with the explanation of the reduction which is characteristic of phenomenological psychology, it becomes immediately clear that there is a great similarity between the two. At first consideration this resemblance is even so striking that one could believe the two to be completely identical. On closer investigation, however, it appears that notwithstanding this similarity there are also fundamental points of difference which come to the fore immediately when we focus attention on the fact that both transcendental phenomenology and phenomenological psychology adopt a completely different attitude in regard to the transcendental problem. Let us try to understand this.

We have seen that the world with its property of "being in and for itself" is as it is, whether or not I happen to be conscious of it. But as soon as this world makes its appearance in consciousness as "the" world, it appears to be related to con-

48. Edmund Husserl, *Cartesianische Meditationen*, pp. 48-63 (7-23).

sciousness. Then it becomes clear to me that whatever exists for me and is accepted by me, exists for me and is accepted in my own conscious life, which, in all consciousness of "the" world and even in all scientific investigation, adheres to itself. In other words, I begin to realize that every showing of truth and being goes on wholly within myself. According to traditional philosophy a great problem lies in this ascertainment. For it is quite understandable that I attain certainties and evidences *within* the domain of my own consciousness; but how can this showing process, going on completely within the immanency of my conscious life, acquire objective significance for "the" world out there? With the advent of phenomenological philosophy it became possible to show that this whole problem is nonsense, since it involves an inconsistency into which modern philosophy had to fall because it did not understand that a transcendental reduction is needed in order to attain that ego by which transcendental questions, that is to say, questions about the possibility of transcendent knowledge, can be asked. However, as soon as we carry out the transcendental reduction and attempt in a systematic self-investigation and as *pure* ego to uncover the ego's whole field of consciousness, it becomes clear that all that exists for the pure ego becomes constituted in itself and that every kind of being has its own form of constitution. This means that transcendence is an immanent characteristic, constituted within the ego, and that every imaginable being, whether immanent or transcendent, falls within the domain of transcendental subjectivity, as the subjectivity that finally constitutes all sense and being.[49]

Be this as it may, it is clear from the preceding discussion that phenomenological psychology is not interested in this transcendental problem and its possible solution and that at any rate it is powerless in the face of this problem. For the purpose of the phenomenological-psychological reduction is not to bring the transcendental subjectivity to light. Phenomenological psychology hopes to expose only the foundations of empirical psychology. It is true that it will never be able to explain these foundations if in its forward development it is interested only in the intramundane beings which manifest themselves in our lived experience. Phenomenological psychology is possible and meaningful only if one is able to perform a determinate reflection in which the lived experiences themselves come to the fore as intentional. This determinate reflection is made possible by a reduction – the phenomenological-psychological reduction – through which the "objective" human ways of behavior studied in empirical psychology are reduced to unities of sense as given in subjectivity as a real psychological entity in the real world. That is why phenomenological psychology remains within the realm of the natural attitude.

However, all this does not alter the fact that there still is a great similarity between the two kinds of reductions. For the transcendental reduction may be considered as a certain further purification of the psychological interest. The psychologist reduces the ordinary valid world to a subjectivity of "souls," which as such are a part of the world which they inhabit. The transcendental phenomenologist reduces the already psychologically purified to the transcendental, to that most general subjectivity which constitutes the world and its "souls." From this point of

49. Edmund Husserl, *Phänomenologische Psychologie*, pp. 331-342.

view it becomes clear that the phenomenological-psychological reduction in its attempt to separate the psychical in its purity from everything which is not psychical has a bearing only on the different modes of man's intentional orientation toward the world reducing them to unities of sense, but does not touch the psychologist's subjectivity itself which, therefore, still is to be understood as a real psychological entity in the real world. Transcendental subjectivity, however, is not a part of this objective world, but that subjective conscious life itself, wherein the world and all its content are constituted for me. Within the realm of the transcendental reduction I, this man, "spiritually" and "bodily" existing in this world, am, therefore, only an appearance unto myself as transcendental ego, so that the "I" which I apprehend here, presupposes a hidden ego to whom the former is "present."

On the other hand, however, transcendental self-experience may at any moment, merely by a change of attitude, be turned back into psychological self-experience. Passing thus from the one to the other attitude we notice a certain "identity" about the ego. What I saw within the psychological reflection as "my" objectification, is within the realm of the transcendental reduction seen as self-objectifying, that is to say, as objectified by the transcendental ego. We have only to recognize that what makes the psychological and transcendental spheres of experience parallel is an "identity" in their significance, and what differentiates them is merely a change of attitude. Under the more stringent reduction the psychological subjectivity is transformed into the transcendental subjectivity, and, eventually, the psychological intersubjectivity into the transcendental intersubjectivity. It is this last which is the last concrete ground whence all that transcends consciousness, including all that is real in the world, and, therefore the phenomenological psychological intersubjectivity, derives the sense of its being.[50]

CONCLUSION

On the basis of this explanation phenomenological psychology can be characterized as an aprioric, eidetic, intuitive, purely descriptive and intentional science of the psychical as such that remains entirely within the natural attitude.[51]

The expression "a priori" refers to the fact that this psychology is focused first of all on that without which the psychical as such cannot be thought of. Only secondarily does it strive to understand psychological facticity, to formulate theories or to find explanations, although our natural attitude makes us interested primarily in these latter aspects. Thus phenomenological psychology is interested primarily in the necessary a priori of every possible empirical psychology.

Intuition and description point to the source of this a priori. The inner intuition reveals generalities which are analyzed, and then what is seen is guided further by the intuition toward general necessities of the intuited situation by means of the

50.*Ibid.*, pp. 343-344.
51.*Ibid.*, p. 46.

eidetic reduction. This method reveals the general chracteristic of intentionality in the realm of the psychical. Now the double polarity involved in the conscious act, which can be expressed as *ego cogito cogitata*, must be considered. Consciousness implies not only the object of conscious acts, but also the ego, the conscious person. Thus psychological investigations acquire typically a twofold aspect and a teleological orientation. From this arises the necessity to investigate descriptively in two directions the systematically interwoven multiplicities of conscious acts which belong essentially to the conscious revelation of the objects of knowledge.

The phenomenological psychologist, however, remains in the natural attitude during these investigations. The transcendental attitude seeks the philosophical, radical, and apodictically evident aspects of our conscious life and in so doing leads to a radically founded philosophy. A psychologist does not wish to leave the non-philosophical, natural, and dogmatic attitude. Yet phenomenological psychology can become a point of departure which will lead to philosophy ultimately, although it can never assume the character of a necessary and founding science. Sciences of the natural attitude are sciences of the world and therefore are sciences which presuppose the world. The eidetic sciences remain also sciences of the world insofar as they seek knowledge of the world.

The pure science of the essence of the psychical life of man and society is *eo ipso* a science of the world. It seeks the apodictic, necessary structure of psychological facts and laws within the realm of facticity. It is only by means of and together with an eidetic science that any empirical science can become a rigorously scientific discipline. In addition to the eidetic psychology an empirical psychology must be maintained which will concern itself with the determinations of the factual as such. The a priori as such provides only a formal framework within which facts insofar as they are thinkable can find a place.

Phenomenological psychology, as the eidetic and aprioric study of the psychical, is distinguished from the traditional empirical psychology in that phenomenological psychology is interested only in the essence of the psychical phenomena and not in facts purely as such, and in that it tries to explain these essences as unities of sense within the realm uncovered by a typical phenomenological-psychological reduction. Phenomenological psychology is distinguishable from transcendental phenomenology since only in philosophy is a transcendental reduction performed.[52]

In conclusion, one further remark is in order. It was noted in the foregoing discussion that in the continuous process of attempting to determine the very essence of phenomenological psychology, especially insofar as it is related to transcendental phenomenology, Husserl's thinking probably reached its final phase about 1928. In his last work, *Krisis*, Husserl qualifies his final description of the meaning of the phenomenological-psychological reduction by stating that even though phenomenological psychology and transcendental philosophy are essentially distinct from each other because of their different reductions, still phenomenological psychology necessarily fades into transcendental philosophy, where it has its ultimate foundation.[53] As Husserl explains it, the psychologist deals with the world as an intersub-

52. *Ibid.*, p. 46-51.
53. Edmund Husserl, *Krisis*, pp. 238-260.

jective communal product of an indefinitely open community of subjects whose conscious lives are interwoven with one another. But the logic of this development demands that the psychologist perform the transcendental reduction, so that it may become clear how he, as a pure transcendental ego, apprehends other egos as similar to him; and so that he may see how he enters into communication with them in order to constitute, by intersubjective cooperation in its diverse forms, the one identical world common to all.

But even if it is true that phenomenological psychology, developed with absolute consistency, turns into transcendental phenomenology, this does not mean that these two sciences are completely identical.[54] The meaning of this statement seems to be only that phenomenological psychology as a theoretical science necessarily strives for and, therefore, also really debouches into transcendental phenomenology. Or to put it in another way, phenomenological psychology understood as separated from a transcendental phenomenological horizon is impossible. In short, there is no psychology that can always remain merely psychology. The endeavor for radicalization, characteristic of every science in one way or another, drives theoretical phenomenological psychology, as it deals with intentional consciousness, into the arms of transcendental phenomenology. In the psychological practice, however, every psychologist must return to the world of our immediate experience, applying there his insights to "real" men in mundane situations.[55]

POSTSCRIPT

FROM TRANSCENDENTAL TO HERMENEUTIC AND EXISTENTIAL PHENOMENOLOGY

Many phenomenologists have been most favorably impressed with Husserl's phenomenology in general and with his ideas in regard to phenomenological psychology in particular. Furthermore, under the influence of Husserl's ideas about phenomenological psychology an important development began to take place in empirical psychology in many countries, particularly in Western Europe. Many psychologists began to use the expression "phenomenological psychology," although very few of them used Husserl's ideas without major modification. In philosophy, too, very few phenomenologists faithfully adhered to Husserl's original ideas. The reason for this must be found mainly in the fact that most phenomenologists, philosophers and scientists alike, had great difficulties with Husserl's transcendental idealism, which in their view is by no means essential to phenomenology as such.[56]

54. *Ibid.*, pp. 261-269.
55. Joseph J. Kockelmans, *Husserl's Phenomenological Psychology*, Chapter VI.
56. *Ibid.*, pp. 314-351; also see the literature quoted there, particularly: Stephan Strasser, *Phenomenology and the Human Sciences*, pp. 245-320; L. Landgrebe, *Philosophie der Gegenwart* (Bonn: Athenäum Verlag, 1952), Chapter 1; *Der Weg der Phänomenologie* (Gütersloh: Mohn, 1963); Walter Biemel, "Husserls Encyclopaedia Britannica Artikel und Heideggers Anmerkun-

This critical attitude in regard to Husserl's idealism, in turn, explains why many philosophers and scientists interested in Husserl's conception finally associated themselves with the hermeneutic, and later also with the existential, interpretations of Husserl's philosophy developed by Heidegger, Sartre, Merleau-Ponty, and others.

In trying to characterize the move from transcendental to hermeneutic and existential phenomenology more adequately, one encounters very great difficulties. Anyone who has, if only superficially, gone through the existing literature knows how arduous and delicate a task it is to arrive at a characterization which would be equally applicable to the philosophical conceptions of Heidegger, Jaspers, Marcel, Sartre, Merleau-Ponty, and others. Thus in trying to describe the events which occurred between 1923 and 1948 we can by no means satisfy ourselves with the remark that during this period the merging of "phenomenology," "hermeneutics," and "existentialism" was brought about. For taken without further specification neither the term "phenomenology" nor the two other terms mentioned have a clear meaning. It seems more useful, therefore, to characterize the new style of philosophy by briefly comparing it to Husserl's original conception of phenomenology. Here again we shall limit ourselves to just those aspects which are immediately relevant for our theme.

After World War I profound dissatisfication developed over the way in which realism dealt with ontological problems; yet the feeling of uneasiness about Husserl's idealism was equally as strong. Husserl's identification of being and being-object called for sharp criticism and the new generation of philosophers certainly could not be persuaded to accept the idea that the existence of an anonymous consciousness, an "impersonal" consciousness as Husserl called it,[57] that is to say a consciousness of nobody, contains the solution for all ontological and epistemological problems.

For Husserl being is being-object-for-consciousness and as such being is constituted by transcendental consciousness. This implies that all modes of being and all regions of beings correspond to different modes of constitution. The final reason for Husserl's adopting the view that being is *a priori* the correlate of consciousness, is to be found in the fact that Husserl wanted to arrive at something genuinely unquestionable which as such can be the ground of all that is. Husserl believed that this absolute root consists in transcendental subjectivity. Now if transcendental subjectivity must indeed be the ultimate ground of all that is, then this means that all that is must ultimately be constituted in and by this subjectivity. Thus all being must be dissolved in the subjectivity's consciousness of it; transcendental subjectivity becomes the only absolute and real being.

Later phenomenologists have offered several major objections to this view. First of all they say, if phenomenology as the process by means of which we let things

---

gen dazu," in *Tijdschrift voor Philosophie*, 12 (1950), pp. 246-289; Paul Ricœur, *Husserl: An Analysis of his Phenomenology*, trans. Edward Ballard and Lester Embree (Evanston: Northwestern University Press, 1967); Hans-Georg Gadamer, *Wahrheit und Methode* (Tübingen: Mohr, 1960); English translation: *Truth and Method* (New York: The Seabury Press, 1975).

57. Edmund Husserl, Ms. 37 IV 26, quoted by Drüe, *op.cit.*, p. 242.

manifest themselves characterizes the standard method of philosophy, then this conception of philosophy presupposes that originally there is something which is not yet manifest and in that sense still hides itself. That which is not yet manifest does not consist in the constitutive activities of a transcendental ego, but in the being of beings (the things themselves) and the totality of all meaning or the world within which these things can appear as meaningful.[58]

Another objection is connected with Husserl's conception of the ego. For Husserl transcendental subjectivity as the ultimate source of all meaning is as such without world. Later phenomenologists claim that the relationship between the human subjectivity and other beings as well as the world belongs to the very *essence* of that subjectivity *as finite*. In other words, where Husserl tries to free the ego from the world and everything which is mundane by means of his reductions, there later phenomenologists claim that the human subjectivity is essentially in-the-world and only as such discloses world. Husserl's pure ego for them is merely an artificial abstraction which is to guarantee the apodicticity of philosophy's understanding of the world. In fact, however, such a conception hampers our understanding of man's own mode of being, his ek-sistence (= his standing-out towards things and the world), and, in the final analysis, of the world itself.

Anyone who adopts the point of view suggested by existential phenomenology obviously must give up the idea of a presuppositionless philosophy and of ever being able to reach apodictic evidence in regard to the world. Among other things this means that philosophy must pay full attention to the fact that being shows itself and hides itself at the same time and, thus, that a certain negativity, a certain experience of "nothingness" always accompanies our experience of being.[59]

Another consequence of this conception of phenomenology is that the starting point of philosophy cannot be found in an epistemology of certain theoretically cognitive activities, but rather in an analytic of man's being as being-in-the-world. For, from the preceding reflection it would appear that the new generation of phenomenologists is trying to find access to genuine ontology by way of an analytic of the "tragedy" of man's ek-sistence which obviously is intrinsically finite, temporal, and historical. To these writers, an existential analytic of man's being is the means of elucidating the general framework in which basic ontological problems can be formulated and philosophically dealt with. But this means that the existential analytic of man's being, in a certain sense at least, has to play the role of classical epistemology.[60] It is even possible in such an analytic to show that certain fundamental epistemological problems are quasi-problems and to explain why and how they could ever have arisen in Descartes' days. This means, furthermore, that it is

58. Martin Heidegger, *Being and Time*, trans. John Macquarrie and Edward Robinson (London: SCM Press, 1962), pp. 49-63; *History of the Concept of Time*, trans. Theodore Kisiel (Bloomington: Indiana University Press, 1985), pp. 13-131; William J. Richardson, *Heidegger: Through Phenomenology to Thought* (The Hague: Martinus Nijhoff, 1963), pp. *viii-xxii*.
59. Joseph J. Kockelmans, "World-Constitution: Reflections on Husserl's Transcendental Idealism," in A.-T. Tymieniecka, ed., *Analecta Husserliana*, vol. I (Dordrecht, Reidel, 1971), pp. 11-35; William J. Richardson, *op. cit.*, pp. 16-24.
60. Martin Heidegger, *Being and Time*, pp. 244-256.

meaningless to try to talk about man in terms of "pure consciousness," "pure ego," and "transcendental subjectivity."

Finally, hermeneutic and existential phenomenology claim that Husserl's postulate of radical presuppositionlessness must be replaced by the basic principle of hermeneutics. Philosophy then becomes identified with a universal ontology based on a hermeneutical interpretation of man's being as being-in-the-world. The Cartesian view to which Husserl subscribes, namely that every philosopher must begin his philosophy by taking his point of departure in something which is apodictically evident, strikes the new generation of thinkers as unreasonable. At the very moment the philosopher begins to reflect he has already engaged himself in the world, society, history, language. It is impossible to eradicate all of this by means of the transcendental reduction. The phenomena, the things themselves, must be accepted by the philosopher the way they really are, but this can be done only by interpreting them from a conception of world which is already there before the philosopher can begin to reflect. But although the philosopher cannot deny that he takes his point of departure in presuppositions, he obviously cannot content himself with accepting them without further investigation. He must try to understand his own presuppositions and must bring that of which they are the presuppositions into a unity with these presuppositions themselves by means of an interpretative unfolding. In other words, the ontological presuppositions are justified by what is ontologically unfolded, clarified, and justified thanks to the hermeneutical interpretation of the meaning as this manifests itself historically.[61]

These few remarks certainly do not adequately characterize the difference between Husserl's transcendental phenomenology and the hermeneutic or existential interpretations of his view by later phenomenologists. If we were to go into greater detail it would become clear, also, that very deep differences of opinion are found among authors such as Marcel, Heidegger, Jaspers, Sartre, and Merleau-Ponty. However, all of this need not occupy us here in that it is not immediately relevant to the main topic of our present investigation. It is of greater importance to reflect for a moment on the implications of the criticism of Husserl's phenomenology for his view on regional ontologies in general and for his phenomenological psychology in particular. But before doing so we must first point out that the leading phenomenologists, such as Heidegger, Jaspers, Sartre, and Merleau-Ponty, all of the problems just mentioned notwithstanding, did not object to Husserl's conception of the meaning and function of "regional ontologies."[62] It is evidently true that in describing these ontologies these authors use a different language, whereas some of them will attribute to these ontologies a more moderate role than Husserl did, but all of this does

61. *Ibid.*, pp. 61-63; cf. Hans-Georg Gadamer, *op. cit.*, pp. 250-289 (pp. 235-274).

62. Cf. Heidegger's draft of the article on phenomenology to be included in the *Encyclopaedia Britannica*, in Edmund Husserl, *Phänomenologische Psychologie*, pp. 256-263, pp. 590-603; Jean-Paul Sartre, *The Emotions: Outline of a Theory*, trans. Bernard Frechtman (New York: Philosophical Library, 1948), Introduction, pp. 1-21, cf. also pp. 92-4; Karl Jaspers, *General Psychopathology*, trans. J. Hoenig and M. Hamilton (Chicago: University of Chicago Press, 1964), pp. 55-154. The fact that Jaspers objects to Husserl's *Wesensschau* does not alter the fact that he, too, maintains the possibility of a descriptive study of psychic phenomena.

not change the fact that in their opinion descriptive sciences are possible and necessary, and that they have a foundational function in regard to the corresponding empirical sciences. Yet this opinion is not shared by all their followers, as we shall see next.

Disregarding the many differences in the phenomenological movement as a whole, we can perhaps say that two important currents in contemporary phenomenology as regards empirical psychology can be distinguished. Those who belong to the first current[63] do not explicitly distinguish a determinate discipline, called "phenomenological psychology," which is essentially different in nature and function from both empirical psychology and from phenomenological philosophy. They usually speak of a psychology based on phenomenological or, eventually, existential philosophy. Thus they only accept *a school or trend* within the realm of empirical psychology which for its philosophical foundations is grounded in phenomenological philosophy. When in this context phenomenological philosophy, which is to give this movement its foundation, is interpreted in terms of Heidegger, Sartre, or Merleau-Ponty the authors most frequently speak of existential psychology, and interpret the term in an analogous way.

The second group of authors[64] holds that phenomenological psychology is indeed *a determinate discipline* which must be distinguished clearly from empirical psychology on the one side and phenomenological philosophy on the other. Following Husserl more closely, these authors claim that phenomenological psychology provides empirical psychology with the necessary foundation for its empirical research by descriptively or interpretatively explaining the human meaning of the phenomena with which it deals in its observations, experiments, tests, and so on.

In this view, particularly promoted by Sartre,[64] the distinction between phenomenological psychology and phenomenological philosophy is not very clear at first sight. Husserl explained this difference by appealing to a distinction between a phenomenological-psychological reduction and his famous transcendental reduction. Since Sartre does not accept the transcendental reduction he has to explain the difference between the two disciplines in another way. He holds that phenomenological psychology is always regressive but that this regression always remains an ideal which in principle can never be materialized systematically. Phenomenological philosophy, on the other hand, is progressive. It can prove that emotions, for instance, are in essence realization of man insofar as he is affection, but it is unable to show that the human reality must necessarily manifest itself in such emotions. That there are certain emotions in man, and only these, manifests clearly the facticity of man's ek-sistence. This facticity necessitates the regular recourse to the empirical, that is, the regression by which phenomenological psychology is characterized and which, in all likelihood, will prevent psychological regression and philosophical progression from ever coming together.[65]

Although the conclusions of Sartre and Merleau-Ponty are quite different from

63. Cf. Stephan Strasser, *Phenomenology and the Human Sciences*, passim.
64. Cf. Jean-Paul Sartre, *The Emotions*, pp. 1-21.
65. *Ibid.*, pp. 93-94.

those of Husserl himself, their view differs less from Husserl's original intention than that of the first current mentioned. Sartre and Merleau-Ponty object to Husserl's conception mainly where this either directly or indirectly is determined by Husserl's transcendental idealism, but they remain faithful to the idea of a phenomenological psychology which occupies a position halfway between empirical psychology and transcendental phenomenology.

In the Dutch school of phenomenological psychology we find the two currents mentioned represented, also. In addition, some psychologists are more oriented in the direction of Husserl's phenomenology, whereas others have been inspired mainly by the works of Sartre and Merleau-Ponty and indirectly also by Heidegger's early publications. It is generally accepted that Buytendijk has been one of the most influential figures in this entire movement. It is thus fitting to turn to his work in order to see how he has tried to understand the significance of Husserl's phenomenology for contemporary psychology.

2

Husserl's Phenomenology and Its Significance for Contemporary Psychology*

F.J.J. BUYTENDIJK

If one wishes to grasp clearly the significance of Husserl's philosophy for contemporary psychology it is surely not enough to establish by a series of citations that in the publications of contemporary psychologists certain theses, thoughts, and concepts of the great German thinker occur in their original formulation. It would be a mistake to regard the rapidly increasing frequency with which we encounter the expressions "phenomenon" and "phenomenology," "intentionality" and "act," "life-world" and "bracketing," and "intuition" in psychological papers as an unambiguous sign of Husserl's direct and decisive influence on psychological thinking and methodological principles.

Nevertheless, the use of new words, in everday life as in the sciences, leads to an altered significance of objects, processes, and the human attitude toward the world. "It is a great error to believe that one can change names without changing the things named,"[1] writes Merleau-Ponty. That is easily understood. Every new word illuminates in a *different* way an experience already designated by a name, or the word introduces to our horizon something which until then had gone unnoticed.

In human society it is easy to show that a word rising up in a historical situation of crisis can transform the intuitive picture of the world, the prevailing system of values and passional behavior. Neither the society nor the individuals composing it asked themselves the exact meaning of words like "liberty, fraternity, and equality" or "blood and soil," yet these constantly repeated slogans profoundly influenced social and political life.

The effect of repetition and imitation in human life is correctly characterized by the usual expression, "fashion." Fashion is not a habit, but a significant behavior, the expression of a value-project on the part of human being as being-in-

* "Husserl's Phenomenology and its Significance for Contemporary Psychology," trans. Daniel O'Connor. From F.J.J. Buytendijk, *Phaenomenologica 2: Husserl et la Pensée Moderne* (The Hague: Nijhoff, 1959), pp. 79-98. Original title, ***Die Bedeutung Husserls für die Psychologie der Gegenwart.*** The translation appeared in ***Readings in Existential Phenomenology***, edited by Nathaniel Lawrence and Daniel O'Connor and published by Prentice-Hall, Inc., 1967. Reprinted by permission of the publishers.

1. M. Merleau-Ponty, *Sens et non-sens* (Paris: Nagel, 1948); English translation by Hubert Dreyfus and Patricia Allen Dreyfus: *Sense and Non-Sense* (Evanston: Northwestern University Press, 1964).

the-world.[2] Missionaries in all epochs have been most aware how profoundly every fashion – even the slightest change in modes of dress or gesture – transforms persons and their thoughts and relationships. Educators are no less convinced of this and they have particularly emphasized the significance of language in the elaboration of a value-structure in the world of children. We can only guess at the influence which the fashionable expression "O.K." is having on contemporary European society.

These few indications may render us attentive to the relations which hold between philosophy and science. These relations have been – and are still, I believe – only *indirect*. Since the time of their autonomous development, the sciences are the result of the painstaking efforts of a continuous community of researchers, bound together by a tradition of *thematization*. The thematization defines what belongs to the world of the scientists, their "horizon of being," which – as Husserl expresses it – has "the character of a unique work or construction developing into infinity." "The scientific world," Husserl continues, "like all worlds shaped by purposes belongs ... to the life-world, just as all men and human societies belong to it."[3] In this "world of simple intuition,"[4] the work of a science clearly based on the objective order, *e.g.*, physics, is conceptually and methodologically determined. No doubt these concepts and methods change with the progress of research but that will not happen through a *fashion*, as for example, through the missionary activity of those philosophers who preach on space, time, and causality. It seems to me thus that Husserl, who already in 1917 complained that *phenomenology* had become a fashionable term, could exempt the natural sciences from this reproach. These sciences have had nothing to do with philosophy since early modern times. If the contemporary physicist would agree with Heisenberg that "the Cartesian distinction between *res cogitans* and *res extensa* is no longer a valid starting point for understanding modern natural science,"[5] still, the insight itself was *not* achieved on the basis of philosophical considerations but through the advance of experimental research.

Modern psychology has a totally different relation to philosophy. Even though in this science, too, contemporary researchers, as a consequence of their own investigations, more and more thoroughly reject Cartesianism and the associated philosophical ideas of modern times, the field of psychology has always been and remains today the missionary field for philosophy. Hence there is in psychology a philosophical fashion, whose influence we may not underestimate. It is often the expression of a discontent or a bad conscience resulting from scientific work. Both conditions may be regarded as preliminary preparations for entering new regions of

2. The reference here is to attempts by, for example, Heidegger and Merleau-Ponty, to give a phenomenological description of the human mode of being, in particular, to Heidegger's *In-der-Welt-sein* and Merleau-Ponty's *l'être-au-monde*. - Ed.

3. Edmund Husserl, *Die Krisis der europäischen Wissenschaften und die transzendentale Phänomenologie, Eine Einleitung in die phänomenologische Philosophie*, herausgegeben von Walter Biemel (Husserliana, Band VI) (The Hague: Martinus Nijhoff, 1954), p. 460.

4. *Ibid.*, p. 125.

5. Werner Heisenberg, *Die Künste im technischen Zeitalter* (Munich: Oldenbourg, 1954), p. 67.

human psychical life, and for renewing the investigation and classification of old problems and long-known facts.

In order to evaluate the scope and depth of the mission of philosophical reflection in psychology, Husserl's in particular, it is requisite to understand that scientific psychology is not and cannot be restricted to those experimental researches, which since Wundt's day have been carried on in university laboratories, but must also include the fields which are called "social psychology," "industrial psychology," "genetic psychology," "clinical psychology," "psychopathology," etc. Studies in these regions are directed toward the knowledge of man and his world. They have given a new direction to the problematic of experimental investigations, as I will show later with some examples.

That is why one cannot – as earlier generations of psychologists attempted to do – separate experimental psychological research as an exact science from all other, merely descriptive, knowledge of man, considering the latter not science but philosophical theory applied to practical life, ultimately to be classified with literature.

If from the whole field of writings in psychology one wished to single out as particularly exact in the sense of the natural sciences the experimental investigations on the relations between physical stimuli and subjective sensations, one would have to share the opinion of E. Straus that: "The powerful, centuries-long influence of Cartesian philosophy – its methodology and metaphysics, with the well-known distinction between two finite substances, *res extensa* and *res cogitans* – is perhaps most clearly demonstrated in the doctrine of sensations."[6] It is precisely this theory – according to which our experiences are caused by objective physical processes outside and inside our body – that constitutes the framework for all psychological thinking in modern times. In the beginning of this period Descartes' opinion was widely shared, that the soul is an independent substance, a knowing being, consciousness, a being in this world, bound up of course with the body but still possessed of particular properties, activities, and contents. But the later scientific psychology replaced this dualistic conception of man and the mystery of interaction with a materialistic monism. Particularly in the first half of the nineteenth century, the psychical was regarded as a product of the brain, a concomitant appearance of physico-chemical processes in the living substance. This representation still dominates medical thinking, and in psychology, too, its influence continues. It is Husserl's great merit to have demonstrated by means of strict philosophical proof the fundamental falsehood and crippling effect of this theory on all psychological research. Husserl based the theoretical foundation of a rejection of Cartesianism not on a new metaphysical speculation but – accepting the suggestion of Brentano – by turning to consciousness as the mode of being proper to human subjectivity, which only becomes *consciousness of something* by an intentional orientation toward the world, that is, by having something in view, by giving meaning.

Husserl's frequently repeated challenge "back to things themselves" has a *two-*

6. Erwin Straus, *The Primary World of Senses: A Vindication of Sensory Experience*, trans. Jacob Needleman (London: The Free Press, 1963), p. 5.

fold meaning for psychology. It requires, on the one hand, an impartial investigation, *i.e.*, one without theoretical presuppositions, without question of the metaphysical reality of what is experienced, based only on immediate experiences. On the other hand, Husserl means that this "back to things themselves" is possible only if a certain *distance* from naïve life and experience is maintained, in order that the meaning-structure of the thing might be read off. And that, in turn, is possible only on two conditions: first, that the psychologist attempt to describe his experiences by carrying through an adequate reflection and, second, that he ask himself on what *mode of intentional orientation* this description is dependent. For classical psychology, oriented on the natural sciences, the scientific perception of appearances was identical with the most exact and analytic determination of the separate characteristics of objects, a perception which turned upon distinguishing and enumerating separate characteristics. Contemporary psychology should ground its findings on the distinction of different *modes* of perception, for example, the categorial intuitions of logical, technical, living, human, aesthetic, or physiognomic meaning-structures of experiences, whether they are simple things and happenings, situations, or interhuman encounters, ways of behaving among one's fellows or among animals.

Husserl has convinced us that analytic perception oriented toward things is only one among many ways of "sensory attention." If this analytic perception, which *excludes* physiognomic representation and qualitative aspects, founds the thematic of a world comprehensible in physical and mathematical terms, one cannot conclude from this that it also constitutes a means of access to truth and reality in the realm of psychical life.

The insight that the physical world, far from being the sole true reality and the universal foundation, is only one of numerous *representations* of the world, projected by a particular thematic – is a liberation for everyday life, which has suffered for three centuries under a tutelage based wholly on Cartesian principles. With the insight into the relativity of natural-scientific knowledge, which in no way diminishes the dignity and value of that knowledge, Husserl has been a liberator for psychology and the human sciences generally. On this basis new trail-blazing principles of method could be worked out for psychology.

Husserl's phenomenology is and remains, for the development of the human condition and consequently for the scientific investigation of all human experiences, modes of behavior, and meaning-structures, as well as for the understanding of the worlds which men inhabit, conceive, remember, anticipate, believe in, or hope for, "the most developed and the most pure form among methods of knowledge – which do not have as their purpose a knowledge of 'brute facts' or the elaboration of hypothetical theories, but rather – place above all else a 'respect for the phenomena' and find their meaning and goal in exhausting the 'contents' of the phenomena."[7] We may understand why Binswanger, to whom we owe this clear formulation of Husserl's significance, once called Cartesianism "the cancer" of psychol-

7. Ludwig Binswanger, *Grundformen und Erkenntnis menschlichen Daseins* (Zurich: Max Niehans, 1942), pp. 642-643.

ogy. This is not the passionate expression of a methodological standpoint but an essential characteristic of the hypertrophy of an idea which has debilitated the vital powers of psychology and spread into all its branches.

It has often been remarked that the psychology of modern times, under the influence of Descartes and his successors – Locke, Berkeley, Hume, Comte, to mention only a few – has only managed to eke out a meager existence. During the nineteenth century, when the experimental investigation of psychical functions under the basic stimulus-response schema became widespread, every philosophical vestige had been apparently excluded from the young healthy science of the psyche. For was it not finally rooted in the same soil in which every science must grow: experience, or more concretely, the exact experience which guarantees laboratory experiments? Has not this scientific psychology made sufficient progress in the last 100 years to indicate a healthy development? And yet Husserl often makes remarks like the following: "One ought to reproach this psychology for being totally blind to the particular nature of psychical life and all the eidetic forms belonging to psychical life, which is a spiritual structure exercising itself intentionally, and a subjectivity constitutive of a spiritual community.[8] What does this reproach mean if not that psychology tends to reject all philosophical reflection on the nature of man, on his existence, on his behavior toward other men, and on the essence of consciousness generally? Exclusive recourse to the pure facts, such as is attempted in empiricism, in no way guarantees a true grasp of them; on the contrary, the meaning, which lies hidden in the facts, remains buried.

Contemporary psychology is influenced on this point by the astonishing activity of North American research institutes. The research is concerned with "facts," defined through "operational notions." Watson's axiom is still maintained: "Psychology as the behaviorist views it is a purely objective experimental branch of natural science." "Its theoretical goal is the prediction and control of behavior."[9]

The application of these principles has had important consequences in many practical areas, as, for example, in the selection of personnel, aptitude testing, and numerous psychosocially or socially conditioned relationships. That is understandable since the facts collected in these investigations and treated statistically are described by means of concepts borrowed from everyday life, that is, borrowed from the world which Husserl calls the *Lebenswelt* ("life-world"). This life-world is, he says, the world taken for granted in ordinary living, which we have come to rely on through experience. "It suffices for everyday practical matters."[10] It is a realm of original evidences, a universe of intuitive primordial vision;[11] but this intuitive evidence appears only in a *naïve* attitude before the life-world.

8. Edmund Husserl, *Vorlesungen über phänomenologische Psychologie*. Summer Semester, 1925. (Husserl-Archives, Louvain: F136, BL 126b) This lecture course has been published in 1962: *Phänomenologische Psychologie*, herausgegeben von Walter Biemel (Husserliana, Band IX) (The Hague: Martinus Nijhoff, 1962). – Ed.

9. J.B. Watson, *Behavior: An Introduction to Comparative Psychology* (New York: Holt, 1914).

10. Edmund Husserl, *Krisis*, p. 126.

11. *Ibid.*, p. 130.

Knowledge of the essence of facts and events begins only when we have established between us and the lived experience of the world a distance far enough so that we can consider them in their "value of ..." A reorientation is meant, in which "the obvious becomes questionable and enigmatic."[12] It is only then that facts, the state of things, become *phenomena* in the full sense of the word, that is, themes of a specific presentation.

In the spirit of Husserl's conception, Heidegger considers a phenomenon "that which proximally and for the most part does *not* show itself at all: it is something that lies *hidden*, in contrast to that which proximally and for the most part does show itself; but at the same time it is something that belongs to what thus shows itself, and it belongs to it so essentially as to constitute its meaning and its ground."[13] In order to understand this definition of the phenomenon, which has become a theme of research in phenomenological psychology, we raise the question: "What is anxiety?"

If I experience anxiety in the life-world, there appears at first a situation, an oppressive feeling and a bodily condition, perhaps also thoughts, memories, expectations, and, in any case reactions, which together make up the experience "I am anxious." In this case I know that I *have* anxiety, but do not know *what* this anxiety *is*. The phenomenon of anxiety remains at first hidden from me in immediate experience. Only in a particular kind of *reflection* can I *decipher* the meaning-structure of the situation, my behavior, the bodily expressions, and my experiences, etc. But that is only possible when everything which belongs to my anxiety is taken up again and rendered once more present through an *act of reflection* which presupposes that by *retention* I maintain the past in the present.[14] What is experienced is "retained"; it remains given to me in the complete meaning-structure, even if only as "no longer there." I can, nevertheless, arrive at an exact description if I thoughtfully consider the anxiety I have experienced as a unity of meanings in its context of external and internal processes without theoretical prejudgments. This *consideration in thought*, by which everything unessential or fortuitous is eliminated and the gaze freed for the essential nature, persisting in every phenomenon of that kind, Husserl calls the "intuition of essence" (*Wesensschau*) or "ideation." It is not a mystical mental faculty but a reflective attitude in which one remains attentive to the meaning-structure and meaning-genesis, hidden in ordinary life but now manifesting themselves.

The *description* of factual behavior in the life-world, for example sexual behavior, yields no psychological knowledge, properly speaking, that is, no *under-*

12. *Ibid.*, p. 184.
13. M. Heidegger, *Being and Time*, trans. John Macquarrie and Edward Robinson (London: SCM Press, 1962), p. 59.
14. Edmund Husserl, *Ideen zu einer reinen Phänomenologie und phänomenologischen Philosophie.* Erstes Buch: *Allgemeine Einführung in die reine Phänomenologie*, herausgegeben von Walter Biemel (Husserliana, Band III) (The Hague: Martinus Nijhoff, 1950). English translation by F. Kersten: *Ideas Pertaining to a Pure Phenomenology and to a Phenomenological Philosophy.* First Book: *General Introduction to a Pure Phenomenology* (The Hague: Martinus Nijhoff, 1982).

standing of human behavior or the human world. The Kinsey Report can be regarded as, at best, a foundation for scientific research. The "facts," which this much-discussed statistical study brings together, are at most as sensational as the announcement that every year in any metropolis so and so many crimes are committed.

It is obviously important in any practical domain – whether pedagogy, sociology, law, politics, or ethics – to know what actually occurs in our life-world and what correlations can be demonstrated in it. But this knowledge is not scientific-theoretical knowledge, not insight, and, above all, not psychology. The policeman who ferrets out a number of exhibitionists in his area, on the basis of a pragmatically usable "operational notion," no more accomplishes scientific work than the psychiatrist who labels a group of patients "compulsive" or "schizophrenic." In both cases one interprets men on the basis of some characteristics or definite symptoms without insight into their *significance*. Only the *genuine understanding* of Husserl's demand "back to things themselves" can make possible the knowledge of the essential meaning-structure of those modes of existence which in the *praxis* of the life-world and practical psychiatry are indicated by the names "exhibitionist," "compulsive," "schizophrenic." Only the demand of "going back, again and again" with the *intention* of intuiting the essence guarantees a deeper psychological knowledge.

An example from the experimental investigation of the perceivable world may clarify this point. In everyday life we speak about the brightness of a color or of colorless light and are readily understood. Should someone ask, however: "*What* is brightness as a pure phenomenon?" no answer is forthcoming from lived experience. Von Hornborstel demonstrated, many years ago in an exact investigation, that it is possible to differentiate exactly the brightness of tones, odors, touch and taste sensations with a generally valid objectivity. This result gave the *occasion* for a "turning back" to the actual phenomenon of brightness and it was found that this characteristic of optical impressions in ordinary experience is a *categorical model of experience*, not only for all sensations, but also for feelings, representations, memories, expectations, moods and the physiognomy of objects, situations, and other persons. This example shows how the extension of the phenomenological method in psychology can accompany the positive progress of this discipline.

It was particularly under the influence of *Gestalt psychology* that interest in the experimental analysis of perceptions, modes of behavior, phenomena of expression, thought-processes, etc. was turned toward the meaning-content of phenomena, initially without any question of a causal psychophysical explanation and without any attempt to dissolve the Gestalt-like relationships of immediate experience into elementary processes or functions. Gestalt psychology, through its rejection of the doctrine of isolated sensations, the theory of association, and all atomistic interpretations of the so-called "contents of consciousness," strongly promoted the interest of psychologists for the ideas of Husserl and his students.

Nevertheless, one would be doing an injustice to the psychologists of the preceding century, the founders of experimental psychology, if one attempted to reduce their insights to a primitive physicalism. As a matter of fact, Wilhelm Wundt had already clearly grasped theoretically the fundamental opposition of the natural sciences and psychology. In his *Grundriss der Psychologie* he wrote: "The mode of knowledge proper to psychology is of an immediate and intuitive nature, inasmuch as it ... studies the content of experience in its total reality, the representations relative to objects *with all their inherent subjective tonalities...*" "Psychology can exhibit the relationship of the contents of experience, as it is truly given to the subject, only if it refrains completely from using those abstractions and hypothetical concepts elaborated by the natural sciences." In Wundt's opinion, "Psychology ought to be considered the stricter empirical science in consequence of the particular character of its tasks." "Just as psychology is the empirical science which *completes* the natural sciences and *lays the foundations* for the human sciences, so it is, for that reason, the preparation for philosophy."[15]

The decision in Wundt's Institute to limit the empirical to experiences in the experimental situation narrowed the research to such an extent that "the total reality of experiences together with their inherent subjective tonalities" was completely disregarded. In spite of this, every psychologist is (as MacLeod remarked),[16] a phenomenologist in certain phases of his work, and so one can also find in the writings of the second half of the nineteenth century important tendencies toward a phenomenological description of experience. Tendencies! Nothing more!

The appropriate response to the question of the *quiddity*, that is, the essence of a "thing" – whether something experienced through the senses, like brightness, color, space, time, or an "emotion" like anxiety or love – was not forthcoming because the paradox was not understood that the obvious features of a life-world which we interpret linguistically "must become questionable and enigmatic."[17] Precisely the *obvious traits* of naïve experience and the modalities of the given must – so Husserl taught – become the theme of scientific investigation. "What was lacking ... was to exhibit consciousness correctly in systematic fashion as consciousness *of* something," Husserl wrote in 1923-24, and he added: "In every vital impulse, the psychical life both human and animal is consciousness *of* this or that. As a whole, consciousness can be characterized as a unitary and continuous flux, always taking on new forms whether of representation, communication, feeling, effort, or action...." "How could psychology arrive on the right path without penetrating to a systematic elementary analysis of consciousness as consciousness *of* something. This characteristic is, so to speak, the *ABC* of psychical life."[18]

Husserl continued to distinguish with increasing clarity the theoretical and methodological foundations of psychology, showing thereby how man in the life-

15. My attention was drawn to this quotation by an essay of L. Van Haecht in *Nederlandsch Tijdschrift voor Psychologie*, 6(1951), p. 41.

16. R.B. MacLeod, "Phenomenological Approach to Social Psychology," in *Psychological Review*, 54(1947), p. 193.

17. Edmund Husserl, *Krisis*, p. 184 & 213.

18. Edmund Husserl, *Erste Philosophie*. Erster Teil: *Kritische Ideengeschichte* (1923-1924), herausgegeben von Rudolf Boehm (Husserliana, Band VIII) (The Hague: Martinus Nijhoff, 1962), p. 53.

world "naïvely" experiences himself in his intentional-*real* relations to *real* objects.[19]

Human modes of behavior, active or passive, that is, the real intentionalities, remain always the *point of departure* for psychological investigation. The investigation itself begins when one is free to look for "the pure inner life considered as the source of all significance and valuation." In experimental psychology, therefore, one should place in the center of attention the *mode* in which the subject relates himself to the environment, that is, *how* he is aware of the situation, *how* he understands the phenomena, *what* they mean to him and *how* he values them. Of course, that is possible only if the experiment is not reduced to the stimulus-response schema, only if the subject finds himself in a situation which is constituted through an active meaning-giving orientation, realizable in the experiment. The situation can be simple and easy to analyze experimentally. Just the presentation of visible phenomena – a schematic drawing, a geometrical figure, a picture-puzzle, a spatial arrangement, an optical illusion, the movement of a spot of light – suffices in order to study the intentional meaning-giving, the genesis of meaning and its relationship to experience, sometimes with bodily conditions and personal affective or characterological factors. *What* the subject sees and how he sees it must be analyzed in experience. The real or imaginary *variation of the given*, as Husserl taught,[20] is the best means for clarifying the intentional project of the meaning-structure. A few examples may illustrate this point.

1. A skin irritation can have the meaning of touching an object or being touched by it,[21] and experimental analysis permits us to know not only the external conditions of this attribution of meaning but also its dependence on the possible or real orientation to the I in prepredicative experience. Just the same tactile irritation in the finger *signifies* something different when I touch something while cautiously exploring in the dark, and when I am unexpectedly touched. This holds already for animals. A blind cuttlefish, when touched on one of the extended arms, reacts by pulling it in, but when the animal itself touches something with the arm, the reaction is a palpation of the object.

2. In the detailed and exact investigations of Michotte[22] on the perception of causal relations, voluminousness, substantiality, and the perception of relative changelessness (so-called "permanence"),[23] the external stimulus-factors for the attribution of meaning to perceived objects were exactly determined. But one can also study the genesis of this attribution of meaning, that is, the modalities of signifying, interpreting, valuing; and this leads to a knowledge of *what*, to the essence, the idea, the *eidos* of a causal relation such as, *e.g.*, pushing something away, forcing something out, taking something along, holding something fast, dragging some-

19. Edmund Husserl, *Krisis*, pp. 247-248.

20. Edmund Husserl, *Phänomenologische Psychologie*, pp. 70-89.

21. F.J.J. Buytendijk, "Toucher et être touché," in *Archives néerlandaises de zoologie*, 10, 2nd suppl., (1953), pp. 34-44.

22. A. Michotte van den Berck, *The Perception of Causality* (New York: Basic Books, 1963).

23. A.C. Sampaio, *La translation des objets comme facteurs de leur permanence phénoménale* (Louvain: Nauwelaerts, 1943).

thing along, etc. -- and to a knowledge of the modalities or "shadings," as Husserl says, in which they appear and can acquire meaning.

3. In an investigation of the mutual influence of figures projected on a wall on their phenomenal mode of movement, it was shown that this influence – called "induction" – is not solely determined, as had been previously supposed, by relations of representational figures and objective characteristics of forms but rather, as Linschoten demonstrated,[24] these factors function exclusively as *motives* for an individually variable projection of meaning onto the optically perceived.

4. In an exact experimental analysis of the relationship of task, success, and failure in human learning[25] J. Nuttin comes to the conclusion that "organism and world, seen from a psychological and functional point of view, are not separate realities." "The psychology of behavior suffers from the isolation of behavior from the soil in which it is rooted: 'man in the world.'" Furthermore, psychology suffers from the consequences of "a traumatic experience, which opposes the conscious or intentional aspect of behavior to an "executive" aspect. Once these two aspects have been artificially separated, psychology raises the now unanswerable questions about their coordination in behavior."

Nuttin demonstrated in his experiments how the "known world," that is, the meaning-structure of the "life-world," need or motive, and "the executive activity," that is, behavior, are essentially only *moments* in the dialectical unity "world-organism."

5. Not only in European psychology, which is more or less in touch with contemporary philosophical thought, but also in the New World where psychology attempts to be free from all philosophical reflection, it has been remarked – though infrequently – how experimental research must take account of the "intentional attribution of meaning" and the unity of man and world. Cantril,[26] who criticized the customary statistical methods and correlations of his countrymen and considered understanding as the fundamental aim of science, demonstrated experimentally the relation between *where* a thing is located in space and *what* its significance is for a subject. A few phenomenological considerations are met with in the many American investigations of "personality in perception"[27] and in the field of so-called "social perception."[28]

6. Finally, I would like to refer to a paper on the "Variations of Intentionality in the Rorschach Test,"[29] because this experimental investigation leads our consideration of Husserl's influence on contemporary psychology beyond laboratory experi-

24. Jan Linschoten, "Experimentelle Untersuchung der sog. induzierten Bewegung," in *Psychologische Forschung*, 1952, p. 24.

25. J. Nuttin, *Tâche réussite et échec* (Louvain: Publications de l'Université, 1953), pp. 467-468.

26. H. Cantril, *The "Why" of Man's Experience* (New York: Macmillan, 1950).

27. H.A. Witkin, et al., *Personality Through Perception* (New York: Harper, 1954).

28. "The Nature of Social Perception," in *Transactions of the New York Academy of Sciences*, Series II, Vol. 19, No. 4 (1948), pp. 142-153.

29. J.M. Kijm, *De varianten der intentionaliteit bij de Rorschach-test.* Dissertation (Nijmegen, 1951).

ments into the field of psychodiagnostics and psycho-pathology. Before we orient ourselves on the application of the phenomenological method in this area, it would be well to cite a few remarks by the author of this paper on the Rorschach Test, Dr. Kijm. The ways in which subjects apply themselves to the Rorschach materials, the modes of intentionality in which they attribute and grasp meaning, are known to depend on their personalities and previous histories as well as on their relations to the investigator, the instructions, and the experimental situation. In order to discover the variations of intentionality in the perception of the chance-figures in the symmetrical inkspots, the author relies on a place in the *Ideas Towards a Pure Phenomenology and Phenomenological Philosophy.*[30] Husserl says in that place, "We can convince ourselves with the help of an example that the neutrality-modification applied to *normal* perception, which posits its objects with unmodified certitude, is the *neutral consciousness of the picture-object*. It is this we find as a component in our ordinary observation of a depicted situation. Let us attempt to make this clear. Consider, for example, Dürer's engraving 'Knight, Death, and Devil.' We distinguish here in the first place the normal perception, of which the correlate is the 'engraved print' as a thing, this print in the portfolio. In the second place, we distinguish the *perceptive* consciousness within which in the black lines of the picture there appear to us the small colorless figures, 'Knight on horseback,' 'Death,' and 'Devil.' In aesthetic perception we do not attend to these as objects; we attend ... to the 'depicted' realities, the flesh-and-blood knight, etc." Starting from Husserl's distinction between normal perception of things and the contemplation of pictures, Dr. Kijm develops the following variations of intentionality in the Rorschach Test:

(1) The understanding of something as figure-ground
(2) The understanding of something as a representation
(3) The physiognomic experience
(4) The understanding of something as a symbol
(5) The understanding of something as a chance pattern

On the basis of these different modes of apprehending the given or different intentional orientations, different behavior-patterns of the subjects are built up, patterns which include a particular, intentional relation to the investigator, for example, a playful or collaborating attitude, an active avoiding of the investigator, or a meditative self-isolation which excludes him. These analyses of intention and behavior in the Rorschach Test are very appropriate for demonstrating Husserl's great significance for psychology. I would like to add, however, that the analyses need to be carried further. Once the psychologist understands the consciousness *of* something as an orientation which attributes and grasps meaning, two questions immediately arise: (1) *what* distinguishes and binds together the modes of seeing – *e.g.* looking in a certain direction, contemplating, looking through, fixating, following with the eyes, etc.; (2) *what* are the fundamental categories of the visible which are actualized in these acts – *e.g.*, the character of thing, event, represented thing, figure, the physiognomic character, the aesthetic character, etc.?

The phenomenological analysis of "seeing" can only succeed on the basis of a

30. Edmund Husserl, *Ideen*, p. 269 (261).

rich experience in a variety of – eventually experimental – situations. It requires a repeated "back to the things themselves," a good selection of exemplary cases and their variations, and "ascent" – as Goethe put it – from simple experiences over the experimentally verified scientific phenomenon to the *pure* phenomenon.[31]

Modern psychology does not rest content with the systematic knowledge of facts and hypothetical explanations but places "respect for the phenomenon" above everything and envisions its goal in exhausting the "meaning-content and meaning-genesis of the phenomenon."

Through the phenomenological orientation in Husserl's sense, psychology finds in its own resources the foundational orientation to human existence, to "bodily existence-in-the-world," to the forms of existential projects and the constitution of the world in the historicity of meaning-giving.

Only in this way will modern psychology find a *new* impetus, *new* themes and problems. As illustrations, I mention the investigation of spatiality and lived-time. Already E. Straus, Minkowski, Binswanger, von Gebsattel, Fischer, and others, on the basis of clinical experiences, have clarified the phenomenology of spatiality and temporality. Much remains to be done, and there is point in citing here Husserl's underlying theme as it is formulated in *Ideas Towards a Pure Phenomenology*:[32] "Here, as everywhere in phenomenology, one must have the courage to accept – instead of reinterpreting – what is really presented in the phenomenon, just as it presents itself and also to describe it *honestly*. All theories must conform to this." What this required "honesty" means for psychology is being more and more understood. The understanding has been facilitated since the psychologist has begun to derive his problematic from the social relations of the "life-world," from art, ethnology, and history – and this in an encounter and participation unencumbered with prejudgments.

In this encounter with existing men and the human aspects of their worlds, phenomenology has discovered the full scope of its task: "to reveal the mystery of the world and of reason." Merleau-Ponty adds: "If phenomenology was a movement before becoming a doctrine or a philosophical system, this was attributable neither to accident, nor to fraudulent intent. It is as painstaking as the works of Balzac, Proust, Valéry, or Cézanne – by reason of the same kind of attentiveness and wonder, the same demand for awareness, the same will to seize the meaning of the world or of history as that meaning comes into being. In this way it merges into the general effort of modern thought,"[33] because – to cite Husserl once more – "the problem which the world poses consciously today, the problem of the deepest essential union of reason and being, *the puzzle of all puzzles*, must become a theme in its own right."[34]

If we understand in this way the affinity of contemporary psychology with art

31. Ludwig Binswanger, *Grundformen*, p. 634.
32. Edmund Husserl, *Ideen*, p. 221 (215).
33. Maurice Merleau-Ponty, *Phenomenology of Perception*, trans. Colin Smith (New York: Humanities Press, 1962), p. XXI.
34. Edmund Husserl, *Krisis*, p. 12.

on the one hand and the *Zeitgeist* on the other, we see that Husserl has definitively indicated as the theme of psychology human existence with its sufferings and its need, its historical self-formation and destiny.

In the first place – and already quite early – psychopathology has demonstrated the fruitful working-out of this theme. Prepared through Scheler's writings on the idols of self-knowledge, resentment, the forms of sympathy, etc., prepared also through Jaspers' use of the distinction between causal and understandable relationships, the phenomenological method proper in the sense of Husserl has gradually become understood and applied by psychiatrists. It is impossible to mention the countless authors and their works here.

I may be allowed, perhaps, to indicate the relation between the widely used psychoanalysis and Husserl's thought. "There are," he writes, "unconscious intentionalities. To these belong the repressed affective experiences of love, humiliation and resentment ... discovered by the new depth-psychology and the types of behavior which are unconsciously motivated by them...."[35]

The Cartesian psychology, which determined Freud's thinking, understood consciousness as something immediately given in everyday life, as a level of the individual man in which perceptions, representations, memories, strivings, wishes, etc., were to be found as thing-like contents with different properties and causal relations. "Even for the 'unconscious," as Fink shows, "there is the semblance of a presentational immediacy in every-day life."[36] The *naïve* theory of consciousness and unconsciousness in psychoanalysis and depth-psychology was a constructed explanation of the experiences of mental processes acquired in the natural attitude of the life-world, in the way they "at first" appear – as real "things" with specifiable properties. Nevertheless it did not escape Freud's perspicacity that every behavior, every experience, has a *meaning*, which at first remains hidden from introspective self-observation but can later show itself in its consequences in the development of personality.

The older psychoanalysis suffers from a fundamental ambiguity, since it accepts, on the one hand, a causal interpretation of intrapsychic processes and, on the other, a meaning-genesis for human behavior. In contemporary psychoanalytic writings, this meaning-genesis is understood as the history of human becoming in its self-constituting relation to an already constituted world. The knowledge which results therefrom must be regarded as an analysis of existence (*Daseinsanalyse*). The deep significance of Husserl's phenomenology for psychiatry appears in the liberation from Cartesianism, in the explicit analytic of intentional consciousness and in a comprehensive knowledge of the intentions and meaning-structures in an individual's mode of existence.

I think one underestimates the influence of phenomenology on psychology if one calls it – with MacLeod – a "fresh look at the world" or an "attitude of disciplined naïvety." The psychiatrist J.H. van den Berg penetrated more deeply into the matter when he urged in his book, *The Phenomenological Approach to Psy-*

35. *Ibid.*, p. 240.
36. *Ibid.*, p. 474. Cf. Eugen Fink, *Beilage XXI.*

chiatry, that psychiatrists should strive to know the physiognomy of things just as it impresses their patients.[37] That is to understand man by starting from his world, but the understanding requires more than a description of that world. It becomes a genuine understanding of man when one has penetrated the intended modalities of the world-project. As Merleau-Ponty says, "To understand is to grasp again the *total* intention" – whether it be the unique mode of being expressed in the properties of a pebble, all the events of a revolution, or all the thoughts of a philosopher.[38]

Once more, Husserl: "In the second half of the nineteenth century the world-view of modern men was determined by the positive sciences. That meant a turning away from the questions which are decisive for a genuine humanity.... Mere factual knowledge makes for factual men.... In our desperate need, this science has nothing to say to us."[39]

At the time Husserl referred to the powerlessness of science to satisfy our "desperate need," meaning a spiritual need, the Second World War had not yet broken out. Since that time – as Guardini has shown us –[40] "our existence has arrived in the vicinity of the absolute decision and its consequences – the highest possibilities and the most extreme dangers."

The end of modern times, in which destiny has placed us, but where we are still responsible in our freedom, forces us to a new formation of our self-understanding. Psychology has a mission to fulfill, the scope and depth of which we can guess but hardly yet measure.

The only thing that is certain is that psychology will fail in its mission if it remains mere factual knowledge. Its contribution to the decisions of our time is determined by an inner bond with philosophical reflection. It was Husserl who laid down permanently the basis for this bond. After him, psychology has developed as a science of human being and a science of the human world; it has brought about a new climate in the life of the spirit and in the human sciences, which perhaps promises a new freedom for the West.

37. Jan H. van den Berg, *The Phenomenological Approach to Psychiatry* (Springfield Ill.: Thomas, 1955).
38. Maurice Merleau-Ponty, *Phenomenology of Perception*, p. XVIII.
39. Edmund Husserl, *Krisis*, pp. 3-4.
40. Romano Guardini, *Das Ende der Neuzeit* (Würzburg: Werkbund-Verlag, 1951), p. 125.

PART II

The Dutch School in Phenomenological Psychology

3

On Human Expression *

HELMUTH PLESSNER

We humans are among those beings which can emit sounds, but our repertoire is both much more extensive than that of all other creatures and richer in modulation. Every sound that an animal utters is determined by its species, its specific character, and by the situation in which it finds itself. Now, you may object and say that a mockingbird or parrot merely echoes sounds when hearing them with no regard to a particular situation, but, though frequently interpreted as dissimulation or imitation, even this is no exception to the rule. *Every* utterance is expressive in the literal sense of the word, ex-pressive; an animal gives vent to its excitement and thus communicates it to its own as well as other kinds of animals, to friends as well as to foes. Consequently, the effect of an utterance is always a manifestation, the function of which varies according to the situation: it may be a luring sound, a warning, a cry of fear, a call to the nest, perhaps even a threat or just showing off. The sound strikes the ear of the animal which hears its own voice. The philosopher Herder found it indispensable to call attention to this fact, for herewith begins vocal articulation. Production seems simultaneously to be the product, production arises as product, production and product are one and the same. For animals, this cyclical process obviously has its distinct limits. Starting by itself, it stimulates others. Through vocal utterance, there is engendered mutual participation. It may well be that other, mute beings possess different channels of communication, in some cases through color, through feeler or antennae, or through movements, such as the abdominal dance of the bees.

THE CAPACITY TO OBJECTIFY AND IMITATE

To call these forms of contact "language" would only breed confusion. Language is spoken, and it represents something. Words and word connections are "associated" with objects as such. As carriers of meaning, spoken words can be detached from the actuality, the state of affairs which they denote; various other carriers of

* "On Human Expression" by Helmuth Plessner appeared in Erwin W. Straus, ed., *Phenomenology: Pure and Applied.* First Lexington Conference, Duquesne University Press, 1964, pp. 63-74. Reprinted by permission of the Publisher.

meaning, such as written marks or other systems of signs can be substituted for them. Spoken words give body to the meaning they convey. Words *mean* something, whereas sounds will, at best, express a condition, serve as a signal or point to something. A scream, a cry of joy, sobbing and groaning, gurgling and grunting, yawning and coughing – all these indicate conditions and situations, yet do not carry nor embody their meanings. Though they may be contagious – think of laughing, weeping, yawning, coughing, for instance – they convey nothing. In contrast to these natural sounds, the signal merely announces; a signal has a definite, precise conventional meaning, one agreed upon in advance: road blocked, railway overpass, etc. The association of the signal with its commonly accepted meaning takes place without benefit of a spoken medium; the signal avoids words. At the same time, words themselves signal meanings; by means of utterances which are independent of a definite situation and are embodied in fixed sound structures available as signs to the speaker, the spoken word refers to things.
sound structures available as signs to the speaker, the spoken word refers to things.

Without this capacity to objectify, there could be neither language nor speech, just as tools could never have been invented. In both areas, in our association with things and with beings related to us, the sense of instrumentality is a *conditio sine qua non* – whether I deal with my hand or with the sounds that I can produce at will. In her comparison between a young chimpanzee and a human being Nadja Koht provides an impressive illustration of this capacity. A child of seven months and a chimpanzee of one year each can produce twenty-three distinct sounds, but in the following months, the human child starts imitating words, and at fifteen months, he uses them to denote objects. His simian counterpart, however, progresses no further, it never imitates.

Formerly, it was customary to explain the difference between man and ape by pointing out its lack of ability to abstract. Experiments in the area of perception, however, have made us cautious, limiting in a significant way the concept of "ability to abstract." Animals, it was recognized, perceive similarties in figuration and also in sequences. But they cannot detach, for example, pointedness, as such, triangularity, or the third in a series, or whatever it may be, from the context in which they have encountered it. A conceptual sense of the general, indeed of generality, is not granted to an animal.

It might not occur to you that this incapacity has anything to do with the simian's lack of imitative ability, for you might not like to acknowledge that monkeys cannot imitate. Now, people have always seen in the monkey the imitator *par excellence*; you even have the familiar expression "to ape", meaning to imitate or mimic, which has the more firmly established this misapprehension. It is an almost ineradicable anthropomorphism, this confusing the ability to imitate, which is a human monopoly, with sympathetic excitability, which is a stimulated reaction, though bearing a great resemblance to genuine imitation, especially since the bodily structure and the behavior of the ape are quite similar to ours. The chimpanzee is suited to a high degree to follow our movements and to mimic us – through its affective structure, its inclination to play, and its marked tendency to participate in and adapt to what it perceives. But the chimpanzee does not imitate human movements by producing them, it merely joins in. The mimicking movements seem to succeed and their success deceives us about their true character. For the most part, the monkey's movements are but responses to ours. In essence, they are not different from the echo barking with which a dog "answers", as it were, the howl of another dog because it is in resonance with its fellow. If you think, for example, of the cheeping of sparrows, or the raucous cackling of starlings when they flock together, you will realize that reciprocal stimulation through sounds is especially favored by the inner relationship between the stimuli, the production of sounds, and the reception of the produced sounds.

As opposed to mere joining in, imitation is only possible where the originator and the reproducer understand the relation as a reciprocal one. This consciousness of reciprocity exists among human beings but not among animals nor between animal and man. At an early age, a child responds to a grimace which I make, to a smile, a wink, to his reflection in a mirror, for he recognizes what he sees as something else, something distinct and separate from himself but directed toward him. An animal is precluded from such recognition. Being able to manage well in a *vis-à-vis* situation evidently does not guarantee knowledge of the trick behind it. Comprehension or understanding of the trick is a much more more difficult task, unattainable not only for animals, but also for human beings until they reach an advanced stage in the development of their abstractive ability.

Acquisition and use of language rest upon imitation as we have defined it, and upon objective representation. Imitating and objectifying proceed from one source, namely, man's capacity to detach himself from himself and to transform himself into something else – in other words, man's remoteness from himself of which he is fully cognizant, this is his eccentric position. Man's upright posture, because of its instability, demands a constant inner surveillance of balance. This leads him to discover his emancipation from his body and its immediate surroundings. He thus is able to distinguish the place where he is, his "here", as opposed to a "there", and to make of a "there" a "here". The capacity to detach, so to speak, his own hand, or the sound which he utters, from himself as a sort of "there" from his "here" is the basic prerequisite for treating hand and voice as instruments, and using them like actual things in the fixed patterns at the disposal of hand and eye. The relegation of everything to the status of things is thus a genuine, a legitimate, aspect of human life, and not in the least a degenerate mode of existence.

The subject matter is realized and represented in the medium of lingual expression, inherited as well as invented, made as well as discovered. The means of expression is detached from the subject matter because, in being articulated, it communicates not only what is said about the matter, but also what is left unsaid. In this we see the great difference between verbal meaning and mere signal or label. It is then that we gain an awareness that speech, though it covers and enfolds the object like a garment, is, at the same time, the skeleton that gives it configuration and character. Speech articulates, dismembers, does violence to the undivided thing, the object itself, but it conforms to it after all, lets it appear, discloses what as garment it seemed to conceal.

We have pointed out that speech has the trait of indirect directness. We might, indeed, go further: speech is the very expression of indirect directness. As such, it is intermediate between the seizing, grasping, forming hand, the organ of distance, as well as the organ which bridges distance; *and* the eye as the organ of immediate presentation. Speech does not merely mark the middle between these two functions, but fuses them into a new function, impossible for either alone. For language, too, is an act of grasping. It makes a matter visible and evident. It is hand and eye in one. Metaphor, which means a carrying over from one spot to another, a transference, is the peculiar achievement of language. For speech transfers; it takes the place of something else; it is the representative go-between, both medium and mediator in the unstable, ambivalent intercourse between man and the world. Being virtually an additional organ, speech, though it certainly does not render the physical organs disposable, relieves them of a great burden. Like a tool which takes over specialized functions, speech assumes the responsibility of vitalizing man's contact with the world. Speech is a means of economy, of thrift, through its "elimination of the organs" as Alberg once described it; that is to say, it is not man who is economized, but through man economies are made; it is a substitute for physical work which has no longer to be performed, a planned action and, at the same time, an institution with established rules, rendering superfluous individual compacts and assuring understanding at its own level as a matter of course.

As a *milieu externe*, speech has a stabilizing and socializing function, for which the dovetailing of a pre-established order and individual effort is characteristic: a person grows into a language structure; he must accommodate himself to its rules, but at the same time, he must master these rules, though he will never really finish learning them. Just as, in walking, our body makes possible at least an unstable equilibrium, so the body of speech (*lingua*), too, gives us the chance to attain an, albeit, unsteady poise in relation to our surroundings. In speaking, I am like my partner, an interchangeable with him, because a reciprocity of perspectives, the dependence of one speaker on the other, is the primary condition of any linguistic structure. I can experience myself as the center of an internal sphere enclosed in my body like a box, opaque to others; I can experience myself as a captive of my consciousness, closed off by an insurmountable and impenetrable horizon which migrates with my sensual impressions and my actions; yet I am liberated from this self-immanence by speaking. In speaking, there is no such thing as a *solus ipse*. Immanence is not eliminated as a possibility, but, when confronted by spoken language, is reduced to a mere aspect.

In the unfolding of a *milieu externe* in which speaker and listener are mutually dependent components, communication on the basis of sound offers certain conveniences – if we had to talk with our hands, we would not be able to do much else. The fact that the voice comes to the relief of the other organs has its basis in the circumstance that lingual articulation is a circular activity: the sound uttered returns to the ear. Nowhere else can the production of a product be so easily perceived; compare it with the work of our hands which congeals into an object estranged from us, or a gesture, which though it transmits contact, vanishes itself. The production of speech, however, is a creation of a medium. Furthermore, voice and mouth possess possibilities of nuance which can be just as much subject to a state of excitement as is conscious articulation. Indeed, many languages associate pitch with meaning. Here we find speaking and singing close neighbors; realms of music that are far removed from the song preserve this neighborly connection – a connection of internal language-gesture.

Imitation and objectification, vital to the acquisition and exercise of language, have the same human root: the feeling for reciprocity of perspectives between my corporeal existence and the existence of another. One part of man's body is especially fitted for this reflexive relationship, the portion that plays a predominant role by reason of man's erect posture and the frontal alignment of his eye: I mean, of course, the face. In the face man expresses himself in an immediate way; the face is, within limits, the zone where the entirety of personal existence is mirrored; including the factor of being seen. As the exposed front which we, with eyes and voice, constantly turn towards our neighbor, it is accentuated and understood contrapunctual to the other parts of the body concealed by clothing. Its mimicking physiognomic transparency is focalized in the eye, the encompassing function of which, to others, appears as a viewing one, as a look. If you think it over, you will realize that the verb "to look" contains both sides of this reciprocity. A glance mediates agreement and understanding between two human beings. In the anthropoids, the various cephalic features are so subordinated to the jutting out of the snout that a play of features cannot take place. Only in man is the facial anatomy such as to free the face. With the placement of forehead, nose, mouth and chin on a frontal plane, reciprocity and the structural principle of verticality come to fulfillment.

LAUGHTER AND TEARS

"Laughter is a rudiment of simian existence; it is an ugly and shameful cacophony. If somebody tickles me under the chin, laughter is ousted somewhere from my body" – this is what Knut Hamsun writes about laughter. It is probably only out of respect for the painful occasions giving rise to tears that he does not include weeping in his sweeping and crushing condemnation. It is true that laughter and tears do not fit well into the image of a self-disciplined, eloquent human being, a man who is constantly trying to go beyond his mere corporeal existence and who, in the words of the Bible, is ashamed of his nakedness. If we, however, review carefully the entire scale of situations to which man reacts with laughter or tears, and to which he evidently cannot react in any other way, then these responses no longer seem so unsuitable, and they lose the taint of being less than human, as it were. They are appropriate to man, even though there are times when it does not seem becoming to give free play to them.

Laughter and tears are responses to limitations imposed upon our behavior. They are manifestations of a certain inability. This ineptitude, it is granted, cannot simply be summed up in the countless, unsystematic failures, large and small, which vex our lives. Laughter and tears are rather of a fundamental nature and are intimately connected with the structure of human behavior as such.

How are laughter and tears stimulated? By anything which is repugnant to man's behavioral structure, not because man's means are inadequate to subdue it, but because the nature of the situation prevents the chance of any behavior whatsoever. By laughing and weeping, man bears witness to the fact that he is at the limits of possible behavior. The experience of being at one's wits' ends, so to speak, certainly has not the same character as an avowal of frustration which is the product of deliberation or trial and error thinking. Indeed in order to explain bubbling laughter or rising tears, we must realize that these prelingual forms of uttering are manifestations of the limits of human behavior, an embodiment, as it were, of the estrangement from one's own body, whereby such embodiment still serves as an instrument in the formation of behavior itself.

According to the usual notion, laughter belongs to the bright and cheerful side of our feelings and tears to the dreary and sad side. Indeed, it is commonly supposed that feelings are always at work and are that to which laughter and tears give vent. In a very vague sense, laughing and weeping may be termed expressive movements, but they should never be confused with genuine mimetic movements, through which affective emotions are discharged, as, for example, in patterns of fear, horror, rage, anger, joy, envy, embarrassment – which may just as well be observed in some of the more highly developed animals. Laughing and weeping can also play a part in discharging emotions, but their participation is scarcely unequivocal. We can laugh or weep for joy, and anger or embarrassment can stimulate either response. But then we are overpowered by the emotional utterance. The language of the patterns which make up our affective mechanisms is not directly at our disposal, it is rather a transparent covering for our psychic constitution which colors it and is reflected in it. Laughter and tears, on the other hand, are completely opaque. The gurgling, gutteral bleating produced by diaphragm and respiration, a lump in the throat, jerky inhalation or the flow of tears, none of these can claim to be colored by or reflected in the frame of mind or motivated by a particular humor. For this reason, they cannot be simply understood as expressive movements. The fact that we are overpowered by them indicates that their function is somehow associated with an obvious interference in behavioral formation.

Human behavior always corresponds to particular conditions which grant a person distance from things and situations as well as from himself. Some sort of order must rule, an order in which and with which matters begin and end. This self-contained order sustains the use of words, social intercourse, and all planned action. At the point where this order as such vanishes, not merely yielding to some sort of disorder which can be rectified, in other words, the moment at which its counterpart emerges, action is blocked, speech is impeded by this shock. If we remain far enough removed from the matter, then the situation amuses us, we find it comical or funny, and we laugh. If we are personally affected by the situation, hence deprived of all distance from the matter, then we are touched or pained, we suffer sorrow or deep emotion, we weep.

Comedy and wit both display a clash of order: improbable situation, ambiguity, paradox or absurdity. Perhaps it is a person who looks like a scarecrow or acts as if he were on strings like a marionette; perhaps taboos are jostled or a person finds himself in an inextricable predicament. Sometimes it is a word or sentence with a hidden meaning, or perhaps a dumb-founding argument, the logic of which proves its very opposite. We might be made to laugh by the exaggeration of caricature or again by the disparity between understatement and the thing it describes. In bracketing everyday situations with satire, irony or humor, we unhinge the world, either emphatically or covertly. Self-irony and humor involve the observer himself in this relativizing process and thus they assure even more definitely his distance from himself. Whether this distance is always accompanied by a feeling of superiority is another question. It is sure, however, that it makes us conscious of a certain freedom and relief, as if the accustomed order of conditions which encircles us, the corset of our lives, as it were, had suddenly released its pressure.

The other situation when behavior is precluded arises when we have no avenue of approach to order and our distance to a matter is abrogated, whether this distance be to myself, to other persons, or to other things. This happens, for example, when we are seized by grief, filled with sorrow or overpowered by joy or deeply stirred by beauty or grandeur. The conditions giving rise to weeping stretch over a broad range: from being the victim of wretched circumstances to being the vessel of the most sublime emotion. We surrender ourselves to these conditions and let ourselves go, for, when we confront them directly, the proportions of our existence are distorted and finally lost. This can be shaming, depressing, irritating or enrapturing. The cause itself, whether trivial or forceful, painful or pleasant, is crucial only because it hits us and is beyond our normal scope. This lack of proportion is not meant in a relative or practical sense; rather, it should be understood as an absolute and completely detached incongruity, as when we are seized by something – to this man responds with tears.

Laughter and tears are symptomatic of situations in which embodiment, the process by which human action is formed and regulated, is prevented. They represent reactions gone astray in a meaningful way, reactions to the impossibility of securing that critical relationship of a person to his body, upon which behavior depends. A person's self-control is associated with and supported by a certain pattern of relations. When this pattern breaks down, when it is altered unexpectedly, and the person loses his self-control, then we see most clearly that man is a being at a distance from himself. Such a sudden breakdown can only befall a being with an eccentric point of reference. Therefore, animals can neither laugh nor weep. Only man has the apex from which he can let himself fall. Only man, knowing the sense of a matter, recognizes that it can have a double sense and that it can be nonsense; only he is cognizant that there is something which reaches beyond the obvious.

As if to compensate for those undisciplined, explosive and disastrous eruptions, as laughing and weeping are in every sense, those modes into which man is thrown when his indirect relationship to the world is blocked, man has a genuinely mimical mode of expression at his disposal, the *smile*. Among all the various forms of gesture, the smile has the honor of being the least bound to any particular emotion. A smile is the slight relaxation of the face, in which all weak and unpronounced impulses are immediately and involuntarily mirrored: wonder, satisfaction, openness to another, understanding; in relaxing the face offers itself as a field of play. The strong affective emotions and explosive reactions of laughing and weeping, however, carry us away because they overpower us. There is no chance of retaining any distance from one's own face. In smiling, on the other hand, there is a poised relationship to one's own gesture: hence, this gesture can assume the function of a mask; it can exhibit tenderness as well as aggressiveness, openness as well as opaqueness. The smile glides effortlessly out of the sphere of involuntary mimicking gestures into one of conscious and deliberate gesture, and may then appear unfathomable, because it says everything and nothing. In this way, man preserves his distance from himself and from the world, and he is able, by playing with this distance, to demonstrate it. In laughter and tears, man is the victim of his eccentric position; in smiling he gives it expression.

Perhaps you will consider these comparatively infrequent modes of expression marginal appearances, although they doubtlessly belong to the constants of human life. But do not forget that even marginal appearances can reveal much, precisely because they are at the margin. The usual analysis of human nature, if it does not pretend to be a doctrine of being, remains bound, to be sure, by the cultural achievements of speech and by the capacity to create in art, technique, cult or society. Similarly, the usual analysis is fettered by those most essential questions pertaining to life and death, security, exchange and social order. But people forget to ask about the roots of these problems and about the possibilities given in the structure of human existence for countering them. The scheme of abilities and drives is only a model constructed in accordance with our self-interpretation as historically encumbered beings. But the historical conscience warns us against such a model. We are obliged to establish the *conditio humana* below its historic sphere of influence, i.e., prior to history. Hence we must turn to those modes of existence which remain invariably constant in the face of all interpretations; these modes we have called embodiment.

4

The Human Body and the Significance of Human Movement*

A Phenomenological Study

J.H. VAN DEN BERG

An attempt is made in this study to submit human movement to an analysis, so that an answer may be found to the question as to the place and the nature of the area from which human movement is directed or – in other words – from which it derives its *significance*.

The very putting of the question points to the fact that human movement is not considered here as a mechanical, blundering process, but as a significant performance, i.e., as a performance having its broad foundation in the whole of human existence.

This analysis will be of a phenomenological character. This does not mean that human movement will be submitted to a close and careful psychological examination (as aimed at by the phenomenology of Jaspers), but that we shall endeavor to find our way from the amazing harmony and discord of man and world to one of the ways of demonstrating this harmony and discord: human movement. It will be clear to the reader that the technical term "Phenomenology" is here taken in the sense of Husserl, Heidegger, Sartre, and Merleau-Ponty.

Since until recently every consideration of the question of the significance of human movement was defeated by the supposedly bipartiteness in man of a material body and an immaterial subjectivity, it seemed necessary to me to demonstrate in a first section the recent development of the subject-conception and then, in a second section extensively to enter into modern phenomenological views about the human body.

It will then appear that the above-mentioned question can be answered without any considerable difficulty.

This study was appreciably promoted by the appearance of the latest, highly important, book by Professor F.J.J. Buytendijk, at present professor of psychology at the University of Utrecht, on the *General Theory of the Human Carriage and Movement*, which up till now has only been obtainable in the Dutch language.[1]

* This essay appeared first in *Philosophy and Phenomenological Research*, 13 (1952), pp. 159-183. Reprinted by permission of the Editor.

1. F.J.J. Buytendijk, *General Theory of Human Carriage and Movement* (Utrecht: Spectrum, 1948) (*Algemene Theorie der menselijke Houding en Beweging* (Antwerpen: Standaard Boekdel, 1948).

As in Professor Buytendijk's book the term movement is here too taken in its widest sense, also encompassing speech.

I. EMIGRATION OF THE SUBJECT

"I will close my eyes, my ears, and shut out all my senses, I will even wipe out from my thoughts all images of material things, or at least, since this can hardly be accomplished, I will consider them false and vain; and thus occupying myself exclusively with my inner self, I will try to make myself even better acquainted and more familiar with myself. I am a thing that thinks, that is to say that doubts, affirms, denies, that knows few things, is ignorant of many, loves, hates, who wills, who does not will, has also imagination and feeling," with these words Descartes[2] describes his *subject* in the opening lines of his third meditation. A thing, thinking, doubting, knowing, loving, hating, willing imagining, and feeling. A *res cogitans*, within the body, separated from the world to which the senses are the misleading entrances, separated also, if possible, from the images presented by the imagination, products too of that same outside world, of whose truth I know nothing and of which I cannot say anything with certainty. An *extra-mundane subject,*[3] to which the world is but indirectly guaranteed through God's goodness alone.

Anyone reading closely the above famous quotation will notice that Descartes himself has had to state that the complete isolation of the subject in exclusive consciousness of self cannot be achieved. Of the wiping out of all images he is obliged to admit that "this can hardly be accomplished." Every rigorous effort to bring this about nevertheless confronts us with a void which has indeed little in common with the experiencing of one's own subjectivity. The isolated subject is, as Buytendijk rightly observes, "a metaphysical point, therefore without dimension, without contents and shape," and – we should be inclined to add – without psychological value: a theoretical abstraction of reality, of which neither psychology nor psychopathology can make anything.

It has surely been this last consideration which has induced Brentano to reintroduce the scholastic idea of the intention (*actus mentis quo tendit in objectum*) into psychology and even to attribute a major significance to it. He considers "every psychical phenomenon characterized by that which the scholastics of the Middle Ages have called the intentional (sometimes also mental) non-existence ("*Inexistenz*") of an object and which we call the direction toward an object (by which we should not understand reality), or the immanent objectivity."[4] It need not surprise us, as Brentano would say, when Descartes observes that the imagining cannot be separated from that which is imagined, no more than remembering from that which is remembered or observation from that which is observed, etc., and that

2. *Méditations* (1641), in *Oeuvres de Descartes* (Paris: Gibert, 1940), Tome II, p. 117.
3. Erwin Straus, *Vom Sinn der Sinne* (Berlin: Springer Verlag, 1935), pp. 8ff.
4. Franz Brentano, *Psychologie vom empirischen Standpunkt* (1874), herausgegeben von Oskar Kraus (Leipzig: Meiner, 1924), pp. 124, 125.

for this simple reason the imagining, remembering, etc., *suggest a direction*, in other words: because every state of consciousness must have an "object," if it is to exist, because every *cogito* implies a *cogitatum*.

Herewith the first step has been taken on the road toward the deliverance of the subject from its narrow unnatural bonds: every quality of the subject is directed and the arrow indicating the direction unquestionably points to the "outside world." The point at issue however is where the end is of the arrow-tip. Not in the outer world itself, as Brentano expressly warns us: by *cogitatum* we do not mean reality, but the immanent object abstracted from it, something like the image which the real object of the outer world represents within one. Hence the title of *non*-existence ("*Inexistenz*").

Husserl was not satisfied with this configuration. If we want to separate the object of the outer world from the immanent "subjective" object, he says, "We get into the difficulty of *two* realities facing one another, whereas only *one* is and can be there. I observe the thing, the natural object, that tree over there in the garden, that and that only is the real object of the observing intention."[5] Husserl therefore extends the arrow, which with Brentano came to a stop at something like the image, to the world itself. In this way every solipsistically inclined view of the subject is wiped out. This becomes quite clear when we see in what way Husserl corrects the intention idea. For Brentano intention was: direction, that is direction from a center (ego) to the outer world, stopping at the image. Husserl examines this road, this direction so to say with a magnifying glass and discovers that with commonplace unproblematic looking, remembering, imagining, etc., there is but very little to be found of the Brentano stages: "if we live so to say in the act in question, if for instance we are absorbed in observantly regarding an occurrence presenting itself or in the play of the imagination or in reading a fairy tale, then nothing is to be perceived any more of the ego as the relevant point in the act performed."[6] The same verdict holds good according to Husserl – and we can but agree with him – with regard to the intention as an arrow between this absent ego and the object of the outer world: while reading, I usually know nothing of a "being directed toward the book"; what there *is*, is the book only. It was therefore bound to happen that a pupil of Husserl's, Specht, exclaimed relieved: "in immediate, unreflected experience nothing has remained of the fact that eyes are needed for

5. Edmund Husserl, *Ideen zu einer reinen Phänomenologie und phänomenologischen Philosophie.* Erstes Buch: *Allgemeine Einführung in die reine Phänomenologie*, herausgegeben von Walter Biemel (Husserliana, Band III) (The Hague: Martinus Nijhoff, 1950), p. 224. English translation by F. Kersten: *Ideas Pertaining to a Pure Phenomenology and to a Phenomenological Philosophy.* First Book: *General Introduction to a Pure Phenomenology* (The Hague: Martinus Nijhoff, 1982), p. 219. This same observation appears already in his *Logische Untersuchungen*, 3 vols. (Halle a.S.: Max Niemeyer, 1913-1921), vol. II, Part I, p. 425. English translation by J.N. Findlay: *Logical Investigations*, 2 vols. (New York: Humanities Press, 1970), vol. II, p. 596.
6. *Logische Untersuchungen*, II, 1, p. 76 (I, 310).

seeing ... there are just houses standing there."[7] We shall see further on that this statement is misleading.

On the whole Husserl agrees with Specht; for him "intentional relation to an object" and "intended object" are identical. And yet we should wholly misunderstand him, if, from all that has just been set forth, we concluded that his subject is to be found in the "outer world." Husserl shrinks from this dictum. When he states that the real object of the observing intention is "that tree over there in the garden" it seems as if he has burned his fingers and must at once redeem the committed error. He does so by emphatically stating that it is of no interest to him whatever whether this tree is really there, he wants to place this and other theses between brackets, he "counts himself out," and retires to a distance to subject the intentionality to a purely philosophical observation. The psychologist however cannot be satisfied with this. He will be grateful to Husserl for this second step on the road to the deliverance of the subject, but as far as he is concerned Husserl and his colleagues will be welcome to the subsequent *psychological* considerations. He wishes to attain a *philosophical* description of the subject.

The third and decisive step was taken by Heidegger. It presents such an entirely new character however that it will be necessary to start from another place in order to be able to repeat his step. This starting place is not the subject but the object, the thing in the outside world.

The object of the physical sciences and of the older psychology has been described by Sartre in a masterly fashion.[8] This object is the bare fact, saturated with being, compact and massive, absolutely independent, unrelated; it is identical with itself and "too much for eternity." This delightful, reliable objectivity is constantly being destroyed by man, more is the pity. His observation is faulty, he constantly makes mistakes. What he sees he confuses with what he saw before or with the figments of his fickle imagination. He turns a thing consisting of cardboard, linen, and pages covered with black marks into a book, a novel; a lock of hair, this bundle of worn threads, into a token of loving memory. A brightly lighted spot in a wood into a ghost and the glass of red liquid into a toast to the health of the guest of honor. He constantly attaches to the object things that are foreign to it and wrong it: "every day in the world we see interpretations applied to sensations without any essential affinity," says Leuret.[9] What is worse, he takes that which he has applied for the object itself; for be careful not to try, when love-making, to take the lock of hair for "what it is"; it will cost you dear. What we see, is not the object, but its signification; this holds equally good for the matter of fact, unemotional, practical attitude: we see, not a thing made of wood, but a hammer, not a piece of white material, but a handkerchief. What we see is exactly what we add to the object, if it

7. W. Specht, "Zur Phänomenologie und Morphologie der pathologischen Wahrnehmungstäuschungen," in *Zeitschrift für Pathopsychologie*, 2 (1914).

8. Jean-Paul Sartre, *L'être et le néant* (Paris: Gallimard, 1943), pp. 30ff. English translation by Hazel E. Barnes: *Being and Nothingness. An Essay on Phenomenological Ontology* (New York: Philosophical Library, 1956), pp. *lxii-lxvii.*

9. F. Leuret, *Sur la folie* (Paris: Crochard, 1834), p. 75.

is to become a *thing* to us. The object, a house, e.g., has no meaning for us in its simple state of being. We cannot give it a name, since with every name a special meaning stands out; it is frozen in meaninglessness. "There emanates a kind of stupor from it to me"[10] and even this is saying too much. What we see in a house is that it is a *house*, we see its inhabitableness. In historical terms: we see "the fringe" of James, the "Sphäre" of Schilder and Kretschmer, the "aurore d'image" of Spaier, the "Erlebnis" of Straus, terms which are intended to indicate that which is added to reality by ourselves. In short: man "projects" constantly what belongs to his sphere into the lucid, transparant, purely existing world of the objects (the word world is too much here) and turns it into an inextricable blend of what *is* and *appears*. "We have of the universe but formless visions, fragmentary, and which we complete with associations of arbitrary ideas, creating dangerous suggestions," Proust[11] splenetically exclaims in the chorus of the psychologists and philosophers. A spleen which is even surpassed, if possible, by Sartre. The views of the latter are so important that we must needs consider them for a moment.

If we define the purely objective as that to which the term *existing* can be applied, then all the rest not belonging to this is *nothing*. There is no word that can express what a cup *is*. We might describe it as a whole consisting of matter constructed in a special way (baked earth) covered with an outer layer of hard matter (glaze). This whole *is*, to us however it is a cup to drink from. What makes the cup a cup is, according to Sartre a "nothing": for the way of handling it, its usefulness is nowhere to be discerned. The unfortunate creature who poisons what is with that "nothing" is man: "man is the creature through whom this 'nothing' comes into the world," he "secretes" the "non-being" and spoils the unproblematical of the objects. He does it thoroughly: "the nothing is at the very heart of the being, in its core like a worm." He renders the world worm-eaten with all his cogitations: with his observation ("to see the world as the world or to turn it into nothing, that is one and the same thing"), with his representation ("to give an image, that is to place an object *outside* the whole of reality, that is to keep reality at a distance, to get rid of it, in a word to deny it") and with his emotion. But after all the world can stand this, is proof against the poisonous flood that humanity pours out upon it: "the non-being is but on the surface of the being" says Sartre, by which he assures us once more that the object wins and must win owing to its sole possession of "being."

Sartre's speculations on "the nothing" do not just drop from the skies. He read Heidegger, although – be it said beforehand – inaccurately. It is important for the present-day psychologist to have made himself familiar with Heidegger's views, since they preclude such misconceptions as Sartre's, misconceptions which – as Sartre convincingly shows us – have momentous psychological consequences.

10. Maurice Merleau-Ponty, *Phénoménologie de la perception* (Paris: Gallimard, 1945), p. 83. English translation by Colin Smith: *Phenomenology of Perception* (London: Routledge and Kegan Paul, 1962), p. 69.

11. Marcel Proust, *A la recherche du temps perdu. A l'ombre des jeunes filles en fleurs* (Paris: Gallimard, 1954).

Heidegger stops at what Sartre considers absolutely self-evident: that the object *is*. It *is* always something, however much I try to divest the cup, standing before me, of all the so-called accidental, man-imposed, which does not essentially belong to the cup; I find myself obliged to keep referring to the cup in words that express more than the cup may be in itself. It is "baked earth surrounded by glaze," I cannot go much further indeed; but baked earth means only baked earth to man, to me, and should we want to drop the adjective "baked," because it implies too much the potter and his oven, even then "earth" is more than ..., than what? "Earth" suggests already all that which, some years ago, Bachelard[12] put into words in such a masterly fashion. The same thing can be said of "the glaze," just as the same holds for all things surrounding us. Even the subtlety of presenting "the cup before me" as a whole of chemical formulas, cannot remove our conviction that the cup thus reduced to "objectivity" *is*, and that a particle of a human word, which, as a result of certain human activities, was turned into a new kind of reality (the chemical formula).

The object is, herein is contained the entire complex of problems of man and his world. The object *is*, this means that the object in its abstraction is immediately and continually transcended into a whole of (human) interpretations, a whole which, eventually, always encompasses the entire world. The cup is earthenware, this means: it implies in principle the potter's clay, the river, the mountains, the potter, the potter's wheel, and the oven; the oven in its turn implies the bricks, the metal, and the fuel. And so forth, the cup encompasses the entire so-called objective world, but at the same time the human world: the potter implies the house in which he lives, eats, and sleeps. The cup is a tea-cup, it suggests the coziness of the time when it is used, it suggests possibly my loneliness in case the guests fail to appear; so it implies on occasion the intimacy of a whole existence, but at the same time the world of things: the chairs round the tables on which others sit or would have sat had they been there, etc. Who could draw the line here between the "objective" and the "human" world? The objective is human, because it *is*, the human is objective, because man interprets himself in the things around him.

The object therefore is never the anonymous, relationless, that is also: the invisible, inaudible, intangible being, with respect to this *being*, it is a non-being, a "nothing." The psychologist will no doubt always regret that Heidegger introduced the word *nothing* here as a *terminus technicus*. Philosophically speaking, he could to be sure point to a long and correlative history of this word. It remains a curious, and in the end to the psychologist, unacceptable contradiction to indicate the transferring of the abstract object to the fullness of human existence by the word *nothing*. In our opinion, therefore, it should be kept out of psychological terminology.

In Rilke's *Laurids Brigge*[13] we find an illustration of the above which should not

12. Gaston Bachelard, *La terre et les rêveries de la volonté* (Paris: Corti, 1948); *La terre et les rêveries du repos* (Paris: Corti, 1948).

13. Rainer Maria Rilke, *Die Aufzeichnungen des Malte Laurids Brigge* (Zürich: Niehans und Rokitansky, 1948), p. 52, 53.

remain unmentioned here. On the blank wall in the Rue St. Jacques, Brigge sees the "impression" of the adjacent house that has been pulled down: the division into rooms, lavatory, staircase, peeled-off wall paper, soot of a chimney, and similar records of an old dilapidated house. What he does see however is much more: "There stood the afternoons and the illnesses and the exhalations and the stale smoke of the years, and sweat breaking out below the shoulders and making one's clothes heavy, the flat smell coming out of mouths and the smell as of fusel oil exuded by perspiring feet.... The sweetish, long smell of neglected babies was there and the smell of fear of children going to school.... One would think I had stood a long time looking at it; but I can swear that I began to run away as soon as I recognized it. I recognize everything here and for this reason it just enters into me: *it is at home in me*." And yet Rilke (Brigge) would have searched for it *within himself* in vain. In order to come to himself thus, he could not do without the wall, while on the other hand it is equally true that the wall "in itself," as object, could never have shown him what he saw, for "the thing ignores us, it rests in itself, hostile and foreign ... an Other Thing, resolutely silent" (Merleau-Ponty, *Op. cit.*, p. 372).

It might be said, not without justice, that the case cited above is too remote from what is ordinary and commonplace to be taken to demonstrate the way in which man realized himself "at the objects." All the same it also holds good for our non-emotional human existence, that as a rule it interprets itself in a whole of meanings that stands out in the things of the world. We find a specially striking case in point with the biologist von Uexkull,[14] who puts the question of the way different people see the same oak tree. For the hunter the oak is a shelter or a cover for game, for the timber merchant a calculable, measurable, saleable object, the young girl sees in that same oak tree an aspect of the romantic landscape. Observing (and imagining, thinking, wishing, etc.: all cogitations mean world), these three persons explain themselves. Never for instance can we tell from the timber merchant that he is a timber merchant. He himself evaporates into a void, when we theoretically take his world away from him. Or rather, when we want to examine the timber merchant himself, e.g., in a psychological laboratory, then this examination will always prove to consist of an exploration of his world. On going to the "object" (in this case: the oak), we find a void there also if we reject every significance. The object in itself cannot be described, or rather: every description of an object proves to consist of an exposition of human world. "For man" says Sartre, "is not coiled up within himself, but is outside himself, always outside, from heaven to earth."[15] If we want to remember our youth, we cannot do without the things belonging to the child's world, bearing exclusively the marks of this youth. The psychology of a child's play, the psychology in general of the child is the psychology of his little room, the garden and the kitten, it is a psychology of the garret with its trunks, boxes, and dark corners, of the cellar with its significant smell and chilliness, of the

14. A.o. in J. von Uexkull und G. Kriszal, *Streifzüge durch die Umwelten von Tieren und Menschen* (Berlin: Springer, 1934), p. 91.
15. Jean-Paul Sartre, "L'homme et les choses," in *Situations*, 6 vols. (Paris: Gallimard, 1947-1965), vol. I, p. 291.

cupboard without tangible boundaries, of the space under the table, where the legs of the parents and of the guests invited to the family feast have their lively play. Thus and in no other way is the psychology of the child a psychology that does not forget the child. And in the same way: solely by leaving the psychically sick child and by putting to one's self the question how the intimacy of his things have changed, a psychopathology becomes possible that does not leave out the child.

Exactly the same holds good for the psychology and the psychopathology of the adult. Whoever wants to get to know a man should leave him as quickly as possible. He is in the last place to be found there where he stands. All the time he silently moves away from himself by expressing himself in the world of things. So one can learn to know another best by traveling with him through a country or by looking at a town with him. One who often shows the same town to different people will be struck by the ever new way in which this town appears in the conversation that is held about the sights during such a walk. These different ways are identical with the people with whom one walks, they are forms of subjectivity. The subject shows itself in the things if at least we dare take these things in their original form, i.e., as they appear, in this case as they appear in the conversation, or in Straus's words, if we take them at their *landscape* value. "Any landscape is a state of soul," says Amiel, who saw clearly the consequences of this observation for psychology: "Every soul has its climate, it is a climate; it has, so to say, its meteorology in the general meteorology of the soul; thus psychology cannot be believed before the physiology of our planet, the science to which we give at present the inadequate name of physics of the globe."[16] Psychology is in the first place – or should be – meteorology, physiology of the elements: earth, air, water, and fire, as psychopathology is in the first place the dogma of catastrophes that afflict the human world. *Psychology* is *cosmology*.[17] This cosmology has found its Copernicus in Bachelard.[18]

Herewith we have arrived at a tentative reply to the question as to the place of the subject, of the "self" that moves. However, before formulating the answer it is advisable to give a recapitulation of the different phases of emigration of the subject in a few points.

1. Descartes: the subject is within the body of man, as in a "closed casket" (Sartre). The only psychological reality is *I think* (imagine, fancy, will, etc.). The inevitable results of this view consist of: the doctrine of the worldless subject, the doctrine of the fixed, stable object, the doctrine of the "sensations" which must be the anonymous building materials of the cognizance of a foreign world, and the doctrine of the projection which pours out over this world matter foreign to the things themselves. Another consequence is the representation of the body as a

16. H.A. Amiel, *Fragments d'un journal intime* (Geneva: Georg, 1901).

17. Cf. Eugène Minkowski, *Vers une cosmologie* (Paris: Aubier, 1906); this is a booklet that cannot be too highly praised.

18. Gaston Bachelard, *La terre et les rêveries de la volonté; La terre et les rêveries du repos; La psychanalyse du feu* (Paris: Gallimard, 1938); *L'eau et les rêves* (Paris: Corti, 1947).

screen between subject and object, as the gateway to the "soul" and as the tool of the subject.

2. Brentano, who corrects Descartes's *I think* into *I think of*, or in order to make clear that this *of* is not used blindly but arrives at a representation: *I think of (this house)*, so gaining the liberation of the subject from the paralyzing "commerce" with itself behind the sealed doors of the senses. With the brackets round "this house."

3. Husserl is not satisfied. He rightly contends that it is this house itself, that house there before me, to which the intention is directed. His correction runs as follows: *I think of this house*, which however on second thought he must correct into *(I think) (of) this house*, because during the being absorbed in thinking of the house there before me there is nothing to be found of an *I* that thinks, nor of an *of*, to which the thinking would be directed. Then however Husserl recoils and puts everything between brackets of the ἐποχή; we might show the final result by means of the formula [(I think) (of) this house].

4. Heidegger and Sartre object to the house "of bricks, wood, and iron." Thinking of a house, I usually think of its habitableness, its intimacy. And should I happen to think of the house as bricks, wood, and iron, then I would be a builder for instance, who at once passes on from these materials toward labor: bricks to be replaced, wood to be painted, etc. Formally I then do the same as the tenant who, returning home from his daily work, thinks of this house "to rest in" or "to pursue his hobbies." The house is, as everything is, always given "*in order to*," i.e., given in the form of a *gerundive* (Sartre[19]). Reduced to a formula: *I think of this house in order to...*, with which the *I think* as well as the *of* and *this house* have "passed," i.e., have not disappeared, but "have been passed over in silence," which we can express in the formula as follows: *(I think) (of) (this house)* in order to.... This *in order* to ... comprizes the house as well as myself. If we now ask where this *in order to...*, which implies the subject, is to be found, then the answer must be: by this house, for there I am, when thinking (observing or imagining), forgetting the house itself, I "am absorbed" in the operative signification of this house.

The anwer to the question as to the place "the self that moves, that acts, expresses and represents" runs therefore as follows: this place is to be found in the world. For that is where "I am," when thinking, feeling, imagining or desiring, I am "with" the things of this world or as language puts it so literally " am absorbed in them."

And yet, we cannot possibly be satisfied with this view. If we fully endorse the final result of the above expositions, even then the phenomenon-in-itself, this man thinking of that house, proves that the definition of the place of the subject cannot possibly be complete. Standing beside him engaged in a conversation with him about that house, then I, looking at him who is thinking, absorbed in that house, am obliged to recognize the other, standing there where he stands, concealed

19. Jean-Paul Sartre, *L'être et le néant*: "Carthage est *delenda* pour les Romains, mais *servanda* pour les Carthaginois." (p. 387; English, p. 323).

behind his eyes or behind his forehead, concealed in the heart of his tangible physical presence. If not, how would Descartes ever have arrived at his conception and with this conception have dominated theoretical psychology until quite recently?

Buytendijk describes a situation from which this relevant problem can be exceptionally well demonstrated. A pedestrian, thirsty and tired from his newly accomplished trip, reaches an inn, where he drops into one of the chairs standing outside. He taps on the table to advise the landlord of his arrival, or rather – for the landlord is only there "transcended" landlord – to obtain the cooling drink that is within the inn. Buytendijk's statement that "the interior of the inn is phenomenally something else than the interiority of the thirsty man" is only partly right. At the moment of the tap on the table, the pedestrian with his chair, his table, his tapping finger, and with everything else that ever was to be found in his inner self, has moved over to the cooling drink within the dark interior of the inn. Phenomenally the inner self of the man and the interior of the inn are *identical*. But the situation changes when we sit down by the side of the parched pedestrian and see him tap. We then see the *parched pedestrian*, we become convinced of an intention guiding his hand, an intention which for the pedestrian himself no doubt originates from the inn, but which for us, the spectators, can be observed in the pedestrian himself: "behind" and in his eager eyes, "behind" and in his testy tapping.

In the following chapter the significance of the phenomenologically unquestionable fact that the subject (or only the other subject, the fellow-man?) is supposed to be *in* the body, that therefore the place of the subject is inseparably connected with the boundaries of the physical body, will be fully dealt with.

II. THE HUMAN BODY

"Our body" says Buytendijk "is that which is most our *own* of all conceivable things, which is least opposed to us, least foreign and so least antagonistic." Any able-bodied person, endowed with animal spirits and full of resiliency, will be able to accept this view in its entirety. And yet it is in curious contrast to the discovery of Wolff[20] that out of ten persons, all in perfect health, only one on the average recognizes his hands out of a small series of photos of which they were told that it would contain the likeness of their own hands. No more does the view fit in with the fact conceivable to almost everyone that our own hand, foot, face, etc., may become curiously strange to us, when we begin to regard the parts of our body attentively, begin to "study" them. Whereas our body is inalienably ours, we do not "recognize" it, when we come to face it in some way or other. We must therefore assume that some fundamental change takes place in our body, which is so absolutely ours, when we *are* it no longer, but see it as a thing among others. This means: body and body make two at the very least; if we want to speak about "the" body, we shall first have to indicate exactly what we mean by it. "The idea 'body'

20. W. Wolff, "Selbstbeurteilung und Fremdbeurteilung," *Psychologische Forschung*, 7 (1932), pp. 251-328.

is by no means one word for one thing," said Marcel,[21] who with this extremely important warning put his finger on *the* weak spot in all pre-phenomenological views about the body. The distinction which Marcel described afterwards had been well-known already for a long time: that between the body that we *are* and the body that ourselves and others have at our disposal, as we *have* it. It appeared that this division into two was insufficient to put into words the special problem of the human body, especially because too little attention was given to the role of the fellow-human in the constitution of the body.

Sartre[22] distinguishes three *dimensions* of the human body. We might describe them as follows.

1. The mountaineer who outlined his plans the day before and discussed his wish to reach a difficult top with his friends, destroys his intentions as soon as he takes his first step on the difficult ground. He no longer thinks of his shoes to which an hour ago he still gave such great attention, he "forgets" the stick that supports him while he climbs and with which he tests the reliability of a rock point, he "ignores his body" which he trained for days beforehand together with an eye to this trip, nor do his thoughts dwell on the closely calculated plan that occupied him so intensely the day before. For only by forgetting, in a certain sense, his plans and his body, will he be able to devote himself to the laborious task that has to be performed. What there still *is*, psychologically speaking, is only the mountain: he is absorbed in its structure, his thoughts are completely given to it. Just because he forgets his body, this body can realize itself as a living body. The body (just as the plan) is realized as *landscape*: the length of the body is demonstrated by the insurmountable steep bits necessitating a roundabout way, the measure of his stride by the nature of the gradient which is just possible or just not possible for him to climb, the size of his foot is proved by the measurements of the projecting points which serve as footholds. The fatigue of his body shows itself in the first instance in the distance or the inaccessibility of the top and in the too steep parts of the way presaging the top. This fatigue shows itself in the first place as the changed aspect of the landscape, as the changed physiognomy of the objects (the rocks, stones, snowfields, the summit), it appears as an ever more obtruding "coefficient of the hostility of the objects" (Bachelard), but not – at least not yet – as a feeling in his muscles, a "sense of effort" (Maine de Biran). The vulnerability of his body becomes clear to him a long time before he falls, in the dangerous incline or as movable objects under his feet; and his pain is present, long before the knock, as pointed rock or sharp stone. Even when he hurts himself it may happen that the landscape takes up his attention so much that the pain is passed on to the "*cave!*" that concretes the terrain; his pain is projected: is the property of the stones, nature of the landscape.

The qualities of the body: its measurements, its ability, its efficiency and vulnerability can only become apparent when the body itself is forgotten, eliminated, passed over in silence for the occupation or for the landscape for whose sake the

21. Gabriel Marcel, *Journal métaphysique* (1927) (Paris: Gallimard, 1935), p. 124.
22. Jean-Paul Sartre, *L'être et le néant*, pp. 368-427 (English: pp. 306-359).

passing is necessary. *It is only the behavior that explains the body,*[23] however long I study my hand, I shall never discover its efficiency in this way. This essential quality is only revealed when I, forgetting my hand, become absorbed in the work the hand does. I may compare right and left hand as long and carefully as I please, I never see in this way the efficiency of the right and the inefficiency of the left, this difference is only revealed when I am going to carpenter or to write, but then again this toll is paid, that I forget both hands.

The eye itself can never teach us that it is the axis of our world; what the eye is, is taught us by the world both visible and invisible, what the ear is by the world of that which produces sound and of that which is silence, etc., or rather – for in reality the senses cannot be put so simply side by side – that the body is a sense appears as accessibility and orientation of the objects of the world. "The senses are the contemporaries of the objects," Sartre[24] concisely formulates.

The same holds for *sexuality*. Psychologically this is not in the first place to be found "in the subject" or "on" his body, but shows itself as world. Nowhere do man and woman differ so fundamentally from one another as in the physiognomy of their world. Sexuality appears in the other, who is met, in the advertisement, the shop window, the reading matter, the landscape of the twilight, the aspect of the street. It appears amongst others clearly from the different value that what is hard, cold, dry possesses on the one side and what is soft, warm and moist on the other.[25] The child, here we must decidedly consider Kunz in the right, is much more primarily and deeply convinced of the difference of sex of his father and mother by "differences in tenderness of his parents' faces" (hard stubbly beard of the father and long soft hair of the mother), than by the results of his infantile sexual "explorations" (Freud). This observation of Kunz's appears to me of primary importance. It opens up a way to a phenomenology of child sexuality in which the harsh handling of the analytical school might be avoided. Such a phenomenological sexuology should find its first task in an exposition of the physiognomy of the world of the boy and the girl, for it is mainly in this physiognomy that the boy is boy and the girl, girl. Just as the grown-up is in the very first place man or woman as to his sexually different world.

Disease too shows itself primarily as a change of the "physiognomy of the world" (Straus). With compulsion-neurosis we are accustomed to speak of a "compulsion," i.e., of a phenomenon that is said to be found with the obsessive compulsive character. Such a patient himself however knows nothing of this compulsion primarily; that which is constantly clear to him consists of a remarkable hostility of the objects around him: "For obsessive-compulsive patients the whole world is filled with

23. Helmuth Plessner, *Lachen und Weinen* (Arnhem: van Loghum Slaterus, 1941), p. 6. English translation by James Spencer Churchill: *Laughing and Crying: A Study of the Limits of Human Behavior* (Evanston: Northwestern University Press, 1970), p. 8.
24. Jean-Paul Sartre, *op. cit.*, p. 382 (318). Herewith Specht's view is criticized at the same time.
25. Cf. for the meaning of the "hardness of man" and the "softness of woman," H.S. Schulz-Hencke's article: "Über Homosexualität," *Zeitschrift für die gesammte Neurologie und Psychiatrie, B.*, 140 (1932), p. 300. Cf. also G. Bachelard's "humidité chaude" (*L'eau et les rêves*, p. 136).

decay."[26] Thus the diabetes insipidus patient recognizes his pathological thirst by the predominant meaning that water has acquired. A patient of von Weizsäcker expresses this almost infectiously: "There is for me a direct communication with water.... I bathe whenever and wherever I can. I often think that it is lovely to have a big jet of water running down one's neck. A brooklet suffices. That is why I want to go to the Black Forest. I am always looking for roads where a brook is flowing through a meadow. Water is purity. I should not at all like to symbolize that. But it must be cold. I do not only drink because I am thirsty; it is lovely in your mouth and throat."[27]

Every pathological change of the body reveals itself originally as a new order in the "external world." The derangements of the cerebrally injured only become comprehensible when, as Goldstein convincingly demonstrated, we stop regarding the patient as an individual having at his disposal an injured body. At the ordinary neurological examination the integrated senile patient displays a set of symptoms which can never tell us in what the derangement of his life consisted which induced his relations to consult a doctor. We only see this when we enter his house and see the rigorous order of the utensils there. Just as we only receive an adequate impression of the senilely disintegrated man, when we see him creating nocturnal disorder in the collected relics of his life. The psychological center of his deranged behavior lies in the appeal emanating from these things in the night, just as the measuredly sedate and careful movements of the integrated senile man are controlled from the small strictly ordered world around him, which has only a limited number of possibilities, a world which he *is*, as has now sufficiently been determined. If this were not generally valid, if, as formerly, we should contend that a man's world was controlled from a center "within him," how could we then understand that his behavior, e.g., depends on the color of the light in which he moves?[28] How could we then ever understand that the knee jerk which occurs in accordance with the rules of the textbook, reverses, i.e., shows a relaxation of the quadriceps femoris instead of a contraction as when the foot is caught by the root of a tree?[29] The *psychological* center of this reflex lies in the world, which does not only furnish all the qualities of the body, but all the way of behavior as well.

If we do not in the very first place define our body as a *world of the gerundiva*, we cannot but consider it an incomprehensible coincidence that Förster's patient suffering from total asomatognosis says of the external world "I cannot recognize anything more"[30] or, like Lhermitte,[31] we shall get into difficulties, when he makes

26. Erwin Straus, "Ein Beitrag zur Pathologie der Zwangserscheinungen," *Monatschrift für Psychiatrie und Neurologie*, 98 (1939), p. 61. ("Für die Zwangskranken ist die ganze Welt erfüllt vom Verwesenden.")

27. Victor von Weizsäcker, *Studien über Pathogenese* (Weisboden: Thieme, 1926²), p. 45.

28. K. Goldstein und O. Rosenthal, "Zum Problem der Wirkung der Farben auf dem Organismus," *Schweizer Archiv für Neurologie und Psychiatrie*, 26 (1930), p. 15.

29. Example by Bethe, cf. Goldstein, *op. cit.*, pp. 111ff.

30. O. Förster, "Ein Fall von elementarer Somatopsychose," *Monatschrift für Psychiatrie und Neurologie*, 14 (1903).

31. L. Lhermitte et E. Tchehrazi, "L'image du moi corporel et ses formations pathologiques," *L'encéphale*, 32 (1937).

the statement that his patient with a serious disturbance in what is called the corporal scheme, "had at the same time lost the faculty of reacting to things, that his entire sense of space had been injured." These and similar observations must remain puzzling when we separate a man from his world.

2. The body that we defined above is the "passed-beyond-in-silence" ("le passé sous silence," Sartre), it shows itself as the world of the subject at the end of its emigration. The *second dimension* of the body comes into being under the eyes of his fellow-man. To describe this body we shall return to the mountaineer, who, absorbed in his task, forgets his body in order to be able to preserve it in all dangers. The only change that we shall introduce into this picture consists in the fact that we assume that the spectacle of the mountaineer in his situation is watched by another, who remains unperceived. Seeing him climb I concentrate on the very thing that the mountaineer himself must forget for the sake of the work he is doing. I see his boots that enable him to keep his footing on steep slopes. I can tell that his body is trained by the adroit movements of his body, legs, and arms. I see the scratches and the bruises that he sustains. *I see his body*, and the whole landscape with which this body contends is centered in this moving living "object." This body is the pole from where the whole mountain and mountaineer is appreciated and understood. Just the thing that is farthest away to the climber, is nearest to me, what must be left by him is for me the center from which comes the significance of all. If I could approach him more closely unnoticed, I could begin to observe and describe this center in detail. I could then see the tense muscles, the drawn eyebrows, the heaving chest, and the sweating face. My knowledge of anatomy and physiology allows me to see the man climbing before my eyes as a *functioning organism*, controlled by the nervous system, which from a center sends its impulses to the contracting muscles. Thus doing I am falling in with the imperative conviction that the body there before me is governed by a center *in* this body; that the muscles and glands are controlled by a principle which this body itself contains. This conviction is independent of my anatomical and physiological knowledge: the living, moving body urges me *originally* to assume an organism, parts of which are distinguishable, parts that are governed by a central principle, i.e., a principle contained within this organism. It is rather the reverse: this conviction makes anatomy and physiology possible: just because I can see the body of the other man as a moving thing is it possible for me to take up tweezers and scalpel and dissect this thing-body. This dissectable thing-body is a derivative of the second dimension of the body. It was especially with this body – not seldom solely with this body – that psychology occupied itself until recently.

Also the mountaineer himself can constitute his body in the second dimension. This happens for instance when he tends the wound in his leg. The wounded place is examined and touched *in order to* cure it, or: *in order to* be able to continue on his way. And it may even occur that he considers his body as thing-body, e.g., when he considers the blisters on his hands as "after all curious behavior" of his body, i.e., regards them as dermatological curiosities. He looks then as it were with the eyes of an extraneous "objective" spectator. It is also with the eyes of another that the girl inspects herself, after finishing attending to her appearance, though the objectivity is different here. Her smile, which she sees in the glass, is destined for other eyes. Through her eyes she allows the other to look at "that face" over there in the glass, which will have to be passed beyond to the meeting, at which the regard of another will play its important part.

3. The *third dimension* of the body comes into being when the mountaineer becomes aware that I am regarding him. Leaving out the improbable, or at any rate rare case that this is a matter of indifference to him, then there are two forms into which this dimension must be divided.

What is most probable and certainly also happens oftenest is that the mountaineer dislikes the regard of the other. He begins to feel hindered, because he knows that the other sees and criticizes just that which he himself must forget in order not to fail in his climbing. He feels vulnerable in an absolutely defenceless domain. It is inevitable that he should to a certain extent adopt the look of the other, now and then or continuously; he does not succeed any more in becoming entirely absorbed in the wall that is to be climbed. He has to a certain extent been deprived of this wall, his world has "flown over toward the other" (Sartre). He miscalculates, begins to make mistakes and will be ashamed of the faultiness of his attempts. There are but few people who can stand it that others look over their shoulder while they are writing a letter and certainly no one can stand it from everyone; regardless of the contents of the written page.

The girl that makes herself up carefully, attends to her eyes and lips, puts on her nylons and adopts an attitude that, as we know, shows nothing but hints at everything, has "*passed beyond*" all these things when she leaves the house and is convinced by the physiognomy of the town that she is well dressed and well groomed. When she passes a group of young men in a quiet street and hears (and "sees") by the shuffling of feet behind her, that everyone turns round to look at her, everything changes entirely. That which she should forget in order to be able to walk – and to flirt if necessary – is now regarded with an "objective" eye. The street before her disappears, the ground under her feet becomes less firm, her carriage becomes constrained, shows in this way even more what she wishes – and does not wish – to hide, she stumbles perhaps and blushes.

The third body appears when the look of the other destroys the "passing beyond."

In my opinion Sartre has no right to leave things at that. There exists – though it occurs less frequently – a look of another that influences the "passing beyond" for good: that makes the world bloom and renders the body straighter and suppler. Innumerable are the declarations of sportsmen that their achievements exceed their expectations owing to the eyes of thousands that are directed on them. There is a loving look that can bestow a *fiat* on my work and at the same time justifies the body that does this work. Just as the caressing hand can for the first time justify that my body is as it is, that the accidental configuration of the veins on the back of my hand appears to be such as it should apparently be (Sartre).

The third body is the body that is constituted while being together with another, the body that falls or is justified in the glance of the other.

4. In the opinion of Van Lennep – which I am glad to adopt here – Sartre's

construction is not sufficient for a phenomenology encompassing the body in all its significations. What there is no room for in his system, is the ubiquitous *primary appreciation* of the body and in particular of the parts of the body (e.g., the calf, legs, hair, hands, nose, eyes), an appreciation which is decidedly not constituted by the supposed or real glance of the other. Everyone lives in peace with the shape and nature of certain parts of his body and in a certain discord with other parts. In between lies the anonymous area which does not count in this primary appreciation. That the glance of the other can be of no importance in this respect may be proved from all those cases where a certain part of the body "primarily narcissistic" is positively appreciated, whereas under the glance of another it acquires an inescapably negative value. An invalid may cherish an immediate, unwarrantable sympathy for his maimed arm, which in the company of others appears a complete deprivation. On the other hand, a part of the body which does not meet with approval immediately, may contribute to the justification of the whole body when making love. As a rule however the two evaluations are equally balanced, but not infrequently we find in that case a certain unconcern with regard to the appreciation of the other.

It is the body of this primary familiarity that is injured by illness, especially malignant disease. In Sartre's system the shock caused by the malignant disease (carcinoma!), the deception and the infidelity into which the primary familiarity is changed by the disease, cannot be placed anywhere. Also the regret at the lost, youthful soundness of the body can only refer to this fourth dimension. Rodin's woman mourning over her withered body ("She who was the beautiful Haulmière") does not in the first place appraise the glance of the other at her wasted beauty, but undergoes it specially as the irreparable loss of a primarily appreciated shape or form.

The signification of the attributes of the body coming under this fourth dimension, such as bath-water, soap, and certainly also clothes (these same attributes have a different signification for the first dimensions) require a study apart.

5. Just as there is a primary appreciation of one's own body, there is a primary affective appreciation of the body or of part of the body of the other. "There are 'unacceptable' hands, false, faithful, primitive hands, etc." say Van Lennep and Strobl in their admirable study on the outward manner of the appearance of man. In all these cases it is possible that a primary appreciation of this part of the body has been expressed. This appreciation or depreciation is then of an entirely different –namely immediate and groundless – character than the constitution of the appreciation expressed in the adjectives, owing to the gesture that the part of the body executes before my eyes: the hands of the other may also *become* unacceptable, e.g., when I see how with this hand he extracts a tablet from a silly kind of medicine container and puts it in his mouth in a repellant way. His hand may become unacceptable through the way in which he rubs the two together (Uriah Heep in Dickens!). The gesture makes the hand that primarily may appear neutral and even acceptable, despicable.

III. THE SIGNIFICANCE OF HUMAN MOVEMENT

When Buytendijk at the beginning of his theory of movement distinguishes process from function, and considers every human movement as a function determined by a sphere of values which furnishes signification to this movement and makes it mean something, then an extremely valuable principle is indicated here, which can protect us against innumerable errors. The movements of the human body are, taken by themselves alone, as mere changes, shiftings, displacements of an organ or a system of organs, not movements any longer. They are only movements in the whole *man + situation*. We saw however that the plus mark represents the most central problematic of theoretical psychology. If it is taken as an algebraic addition, then the inevitable result is that we get caught in the determining of the place "containing" the significance of the movement: we are unavoidably compelled to the localization of this significance in the subject encompassed by the body, whereas every day experience teaches us that the movement is determined by the situation in which man finds himself, that therefore the significance of the movement is supplied "from the outside." We shall now try to formulate an answer to the question regarding the locality of the field that furnishes significance, an answer that does not leave us in the uncertainty indicated here.

For this purpose we begin with the movements of the body described by Sartre as first dimension. This body of the "subject at the end of its emigration" is realized as "utensil," as "domain," in short as *landscape* in the sense of Straus, it defines itself through the chair on which it sits, the pavement on which it walks, and the threshold over which it stumbles. It is exclusively present as the *world of the gerundiva*.

Just as perception, this movement of head and senses, "can only occur at the same place where the object is perceived and *without distance*, is also the warning, beckoning, explanatory, pointing, or groping movement of my hand to be found exclusively there where the "intention meets its terminus": "When I make a sign to a friend to approach me ... I make a sign there where my friend is," says Sartre, and when I perceive afterwards that my friend does not show any mind to accept my invitation, "my gesture of impatience proceeds from this situation without any intermediate thought." We can conclude from this: the field that furnishes significance to my movements, lies there where my friend is, where the indicated house stands, where the root of a tree nearly made me stumble. This is to say: there where phenomenally the movement is affected, is at the same time to be found the domain that causes my movements to be as they are, that determines these movements, gives them a meaning, or makes them significant. Since the mover must forget himself in order to move "over there," that is at the same time the place where the mover is. *Significance*, the *somewhere* where the movement is made and the *self* that moves, all coincide in the landscape, in the "things," but, be it understood, not in the things themselves (the objects), but in the things taken as *gerundiva*.

An essentially new constellation comes into being, when a second person observes the movement of another in a certain situation (body in the second dimension of Sartre: the body of the other as it appears to me, without the other being able to make my glance count in his movement). What I then see is a panorama, grouping itself round the similarly observed movements of the other. These movements are the center whence the landscape receives meaning: the body of the mountaineer exerts itself to the utmost through the long steep slope that it "takes" in its movemens, it performs unbalanced antics because of the threshold over which it stumbles, it shows certain movements of the hips through the physiognomy of the "Saturday night" street disclosing itself with these movements. That is to say: "the body forms itself in anticipation of the aim it serves, it assumes a "shape," ... a shape for doing work, for fighting, for feeling," as well as a "shape for loving," which only awakes at the "similarly directed unfolding of a differently sexed body," as von Gebsattel[32] so delicately expressed it. The flirting girl moves as I see her move, because her world is that of the game of love; the child moves as a child, because his world does not yet know the grown-up order and the use of things. Only in this way are the movements of the other-in-his-situation significant and comprehensible.

Buytendijk and Plessner have perfectly recognized this relationship which holds good for the animal as well: "Behaviour means replying, and we understand the answer, when we observe the animal from the situation that thrusts its question upon him."[33]

In this case therefore the *significance* of the movement lies in the situation in which the man who is observed performs his movements, for from this situation they receive their meaning. The *somewhere* where the moving takes place (the situation as *gerundivum*) is now however no longer identical with the *self* that moves. This *self* is now the mover himself, he there in front of me; it is true he *has* "to pass beyond" always, but for me, the spectator, he is the point of junction of the gerundiva; the center, which before my eyes replies to the appeal issuing from the landscape.

The incongruity which we meet here of the locality of "somewhere" and "self," makes possible a shifting of the field that furnishes significance. We will presently return to this, when we have paused a moment at the consequences of what we have just observed.

The theory of movement which results from the above-described *first* dimension of moving, must consist in an exposition of the wealth of forms of the landscape (situations). Buytendijk has taken the first step in this field of scenic semantics when he compared the movements of herbivora on the one hand and of carnivora and monkeys on the other: the difference only becomes quite clear when we observe that carnivora and monkeys live in "a differentiated domain," which is diametrically different from that of the herbivora. Straus[34] made a similar observation in the field of psychiatry: "Feeling and movement have equally changed for the

32. Victor E. Von Gebsattel, "Süchtiges Verhalten im Gebiet sexueller Verirrungen," *Monatschrift für Psychiatrie und Neurologie*, 82 (1932), p. 113.
33. F.J.J. Buytendijk und H. Plessner, "Die physiologische Erklärung des Verhaltens," *Acta Biotheoretica*, Ser. A, vol. I (1935), p. 169.
34. Erwin Straus, *Vom Sinn der Sinne*, p. 153, note.

catatonic. Everything is already there for the catatonic." The only possibility for us to completely understand the behavior of the catatonic is if we understand the fact that his world has totally changed. Only when the world has been reduced to the rigidity of timelessness can man become motionless, his lack of movement receives its *significance* from the catatonic landscape.

We understand the *dance* only from space transformed by music or by silence. The "past- and futureless space"[35] of the dance, widely differing from our daily world of utility, creates the dance, inspires the dancer with new life that metamorphoses her body.

In the psychology of religion it was Guardini who made a first attempt at describing the liturgic gesture from the world of the faithful.

However valuable all these initiatives may be, psychology – just as psychopathology – is even now confronted with a gigantic task, rich in promise in my opinion, of describing the variety of human landscapes. Only then a great gap in the theory of human movement will be filled.

The incongruity of the locality of the *somewhere* where movement takes place and the *self* that moves in the above described *second dimension of movement*, makes it possible that the place from where movement receives significance is shifted from the landscape to the man who is moving there before my eyes. For, if I should wish to do so, I can see his movements as changes that are fed by a significant intention perceptible in the body itself. The mountaineer "has a plan," he "estimates all his possibilities," he "husbands his strength." In the same way I can tell from the pedestrian himself, i.e., from the movements of his body, whether it is his intention to walk 2 or 20 miles, just so – as was described above – it is possible for me to determine the length of the stage from the physiognomy of the lanscape, which, as we know, is different with the short or the long distance march. In the first case the significance of the movement is *in the intention* of the pedestrian, i.e., *within* his physical body, within the *self* that moves; the second case on the contrary, in the appeal issuing from the landscape, so: in the world, or: in the *somewhere* where the movements are performed.

De Balzac says: "All the women that have gone wrong are remarkable for the exquisite roundness of their movements. If I were the mother of a family, those sacred words of the dancing-master: *Round your elbows*, would make me tremble for my daughters." If we should ask what made De Balzac so apprehensive at the non-angular, rounded movement, then the answer must point to the pliant, accessible, immediate world, which rounded the movements of the fallible women. For he wishes to protect his future daughters against this *world*. The gracious rounded movement alone would never have made him uneasy, what he knows however is that this movement is exclusively constituted by a "dangerous landscape." The significance of this gracious movement is in this landscape. But in contradistinction to this we can without any trouble make an equally correct observation: the significance of the gracious movement is within the person who moves, in her "constitution," her "intention," at any rate *in her*. I deprive myself then it is true of a vision,

35. Cf. Straus, "Die Formen des Räumlichen," *Nervenarzt*, 5 (1930), p. 633.

which after all is alone able to show the movement to its utmost possibilities: the latter observation is therefore not only essentially different, but also poorer than the former.

We should wish to illustrate all this further from *the spoken word*, thus from a representative movement in the sense of Buytendijk.

If we ask about the locality of the domain that provides the spoken word with meaning, then the answer consists of different parts:

1. If I am with the other, speaking and listening, to the thing in question, just as the other, listening and speaking, is absorbed in the world that unites the two of us, then he and I go beyond the sounds produced by our organs of speech for the sake of the thing, visible or understood, to which our words refers. Speaking about "that dilapidated house over there," I am with the other at that house, word and answer originate from the house. Psychologically speaking there is only that house, which in its modus of gerundivum contains the subject of the other and of myself. In other words: the *somewhere* where the movement takes place and the *self* that moves, are there where the word – this representative movement – is furnished with significance. We have already paid attention to this construction, which lies within the first dimension of the body, defined by Sartre, as to the two ways in which movement appears, when the body is taken in its second dimension. These two ways will be described once more under the following points – and now in connection with the spoken word.

2a. When I look at the other, while he speaks to me, I see him primarily as I watched the mountaineer in his field of activity. That is to say: I can let him, the speaker, be constantly announced by the meanings of his world, which, while he speaks, constantly provides him with words. In this case I understand his word from the plan of his world, from the nature of the landscape from which his words are born. The significance of the movements of his mouth, of the sound complexes that fill the room, of the movements of his hands, of his whole speaking body, are, just as described under point 1, provided by the landscape; it is also there that the word happens, i.e., where there is movement; but the *self* that moves is not identical with the place of the *somewhere* where the movement takes place.

2b. Looking at him, the speaker, I can shift the significance of the spoken word from the landscape to the speaker who is standing facing me. The problematic which is contained in every landscape – much could be said of it – then becomes dubiosity of the speaker. The partial or dubious elucidation of the common subject of conversation changes in the obscurity of the intention of the other toward me. The "cryptology of the psychical" which first consisted in the landscape illuminated from one point only – and consequently always partially hidden – becomes the hidden inner-self of the other. His face, which was first (under 2a) the mirror of the landscape that called up his word, then becomes " a façade, a boundary – and a communication wall of his own self towards the other, of the inner-self against the external world."[36] The face is then the mirror of his inner-self, of the heart, this "place of his decisions." Then the *significance* of his movements is transplanted to

36. Helmuth Plessner, *Lachen und Weinen*, p. 49.

the head and the heart of the other, to the centers of the *self* which moves there before me, while the *somewhere* where the movement takes place always remains his world, the "landscape" which he is speaking about. For there *he* dwells, transcending himself; *I* however do not take part in this transcending, but measure the intention of head and heart.

The view which wished to regard speech as the secretion of thought produced in cerebro, is a derivative of this original perception. There is sense in this view in so far as it was born from a phenomenological datum, how for that matter could it ever have been formulated otherwise? The same holds for the different theoretical views on the dialogue. The association – the "analogy conclusion" – and the "Einfühlungs-theory," they all spring from this last way in which speech appears to our eyes. *They retain their validity*.

3. When the other perceives that I do not transcend with him to a common landscape, there arises in him Sartre's third dimension of the body and simultaneously the movement modus correlate to it. He perceives that my look does not aim at the subject (the house over there) via him, but knows that I scan his intention. My look hits him just there where he is unprotected: in his silently transcended body, which is now kept present by my look. Simultaneously he is entirely or partially deprived of the landscape. He becomes uncertain, stumbles over his words (for these are now no longer placed at his disposal by the landscape, but – also for him – flow out of his mouth), or masks his uncertainty by means of a hard, loud voice. The dialogue is wrecked.

The inquiry into the locality of the domain that provides the undermined words with *significance*, is the inquiry into the origin of this undermining. The answer cannot be difficult: this domain is the *glance of the other*, for exclusively through and in that look originates the stumbling, stuttering, making mistakes, blushing, and converting. The *significance* of the movement lies in the look seeking unmasking, which is directed at this moving body itself. The somewhere where the movement takes place also lies in this look for there the person who is regarded looks for a way of escape in vain. The self, which moves, is the person looked at compelled back within the bounds of his own body, looked at and requested to unmask.

We have already pointed out that Sartre's phenomenology of the look must be called one-sided. For to him the look of the other always effects an alienation; his eye robs me of my word, it makes me invariably "coagulate," as if I had been caught in the act. All the examples described in *L'être et le néant* to support this view are all equally misanthropic in character: a man peers through a keyhole at a scene not meant for his eyes and suddenly realizes that his reprehensible behavior is observed; another walks through a deserted street and hears behind him that an unknown person pulls aside the curtains to look silently after him. Sartre's look is the look from behind, the malicious look of an unknown person, the look that causes a shiver from neck to ankle. It would not be difficult to give a long list of examples showing a diametrically different meaning of the look of another. There is a look of understanding, of sympathy, of friendship, and of love. The mere wink may convince us of the reality of a being together not doomed to a Sartrian "échec."

Consequently the eye of a fellow-man may justify my body and my movements. It may impart to me a happiness far exceeding in value any solipsistic satisfaction.

With this appreciation of the look of the other the structure of *significance, somewhere,* and *self* remains the same however. The *significance* of the movements justified by the look lies *in the look*, for it is the eye of the other that justifies my body, that makes my words sincere and my actions transparent. The *somewhere* where the movements take place lies in *this look*. The movements it is true take place in the landscape (the picture I make, while another loving looks on; the flag I hoist under the eyes of the eager spectators), but this landscape no longer concerns my own subjectivity, it has come into being through the instrumentality of the loving or friendly *glance of the other*. The *self* that moves, *I* am, in an entirely different fashion, myself, there in the spot where I stand. Under the encouraging stimulating look I know what I do and *that* I do it; every action that I perform is action of my hand, my arm, my body. The accepting look of the other gives me the almost exceptional right to be myself as a *moving body*.

Now that we have arrived at the end of this investigation, we can conclude that to the question as to the place of the domain furnishing significance to human carriage and movement, not one, but three answers must be formulated. Three domains can be indicated from where human movement receives significance, *viz.: the landscape* (in the sense of Straus), the *inner self*, and the *glance of another*. The theory of human carriage and movement ought therefore to be *divided into three chapters*.

The first chapter comprises the exposition of "the subject at the end of his emigration," i.e., therefore of the human landscape, of the physiognomy of the things in which man daily realizes, and reveals himself. This first chapter consists in the scientifically justified interpretation of the language of the "mute things," it indicates the manifold nature of the appeal that the world makes, makes in such a way, that the silently transcended body replies to it with a certain attitude or with a certain movement.

The second chapter treats of human movement as "visible expression of an invisibly organized unit, which – dwelling "centrally" in head or heart – leads a life of its own, fed by a spate of impressions, which is conducted there by means of the receiving organs."[37] This chapter aims at the exposition of man's inner self; of the secrecy of head and heart, of the mystery that a fellow-man remains and which is the cause that even in utter closeness he is the *other*. It shows in what way this hidden inner self becomes visible in the movements of the body. The chief word in the terminology of this chapter is *function*, and rightly too, for it is a question of reciprocity between two domains: the *functional reciprocity* of man's inner self and the world categorically separated from this inner self. In this chapter, but there only, the use of the words association, intention, projection, instinct, libido, etc., is justified, since these ideas are inseparably connected with the conception of the duality "inner self" and "external world." Within this chapter man *has* a soul and a body, a disclosure and a mystery, a conscious and an unconscious.

37. F.J.J. Buytendijk, *op. cit.*, p. 323.

All the prephenomenological theories about human carriage and movement known up till now start from the just mentioned dual conception. It is an additional, and in my opinion very important, task of this second section to rehabilitate these theories (also the Cartesian one) after a critical analysis of them.

The third section shows the modification which human movement undergoes when it takes place under the regard of another. It does not only comprise all the forms of the alienation of the moving body, but at the same time all the ways in which this body is accepted in its movement by the friendly or loving look of the other. It goes without saying that in this section an important place is due to psychopathology of carriage and movement.

The first and the third section are still waiting for the author who can integrate the extensive material which has meanwhile accumulated.

5

On Falling Asleep*

JAN LINSCHOTEN

> "A flock of sheep that leisurely pass by,
> One after one; the sound of rain, and bees
> Murmuring; the fall of rivers, wind and seas,
> Smooth fields, white sheets of water, and pure sky;
> I have thought of all by turns, and yet do lie
> Sleepless!..."
>
> W. Wordsworth[1]

I. INTRODUCTION

These words of a desperate poet unable to get to sleep plunge us immediately into the midst of the problematic which will interest us here: the question of falling asleep. All of us know of the conflicts and exasperations which surround not being able to fall asleep. We all remember those nights when we tried everything but without success; on the contrary, the more we exerted ourselves the more awake we became. The insomniac tosses and turns in his bed, continually changes position, sighs, squeezes his eyes shut, stops the clock that is two rooms away, puts cotton in his ears, is warm and cold in turns, listens to his heartbeat, tries all the well-known tricks without success – and then in an unguarded moment falls asleep. One day he picks up a textbook on psychology only to discover to his amazement that falling asleep is not dealt with in it. Why is this so? Why is it that so little attention is paid to such an important subject as falling-asleep which either happens each night or fails to happen? According to Kleitman who devoted 600 pages to the subject of "sleep and wakefulness" a special description of the psychic state of one who suffers from insomnia is superfluous as all of us sooner or later go through this torment.[2] If we want to know why we are tormented and how we can fight insomnia, it seems a *conversation with falling asleep* is the only thing left for us. We wish to question this phenomenon and try to understand it in its essential structure.

* "Over het inslapen" appeared originally in Dutch in *Tijdschrift voor Philosophie*, 14(1952), pp. 207-264. Reprinted by permission of the Editor. Translated for this volume by Joseph J. Kockelmans.

1. W. Wordsworth, "To sleep", Sonnet XIV in *The Poetical Works of William Wordsworth*, vol. III (Oxford: The Clarendon Press, 1946), p. 8.
2. N. Kleitman, *Sleep and Wakefulness* (Chicago: The University of Chicago Press, 1939), pp. 380-381. This work contains a very useful bibliography on sleep of over 1,400 items, mainly physiological in orientation.

But how are we to accomplish something like this? The fact that at this moment I am not allowed to fall asleep already begins to characterize the situation. We are unable to place ourselves in the situation we wish to examine – since sleep and thought do not go together. Does this mean that the answer to the problem of falling asleep can never be forthcoming? Is it possible to have a conversation with a partner who precisely silences all conversation? An old German rhyme[3] describes the situation:

Wie man einschläft, möcht ich wissen:	I would like to know how one falls asleep
Immer drück ich mich ins Kissen,	Over and over I press myself into the pillow
Denk dabei: "Jetzt geb ich acht."	And thereby think: "Now I will pay attention."
Doch eh' ich mich recht besonnen,	But before I have really reflected,
Hat der Morgen schon begonnen,	It is already morning,
Bin schon wieder aufgewacht.	And I have again awakened.

And yet one thing is certain: even though we do not know anything while we are asleep, we nevertheless know after we have slept that we *did* sleep and that we *did* know things before we fell asleep, that this knowing-before-we-fell-asleep gradually evaporated while we were falling asleep until we sank into a deep sleep about which we have no positive knowledge at all.

Thus although a proper reflection is excluded while we are falling asleep our consciousness of falling asleep is not completely excluded from analysis. True, this fact limits our reflection, and even in a double sense. First of all, spontaneous reflection disappears during our falling asleep, as is correctly expressed in the rhyme above. Secondly, in view of the fading-out character of our falling asleep it is impossible to form a clear and sharp idea of this consciousness in our theoretical reflection. Furthermore the state of unconsciousness characteristic of sleep cannot be penetrated by theoretical reflection.[4] And yet we hope to show in what follows that an analysis of falling asleep is not meaningless.

We can distinguish between three 'levels' of consciousness: the realm of *reflection*, the realm of our *non-reflexive experiencing and acting*, and the realm of our *life as such*. We conceive of sleep as the completely unreflective and experienceless "life without further qualification"; and of falling asleep as the silencing of reflection and the return of experience to its ground. Falling asleep is accessible to our investigation not only insofar as it is *still* reflexive, but also insofar as it can be made transparent in its essential structure by thought; on the other hand, we can evoke the experience found in our falling asleep with regard to its disposition, color, and climate, in a word, as an experience sphere in order to make it explicit in its essential characteristics. All human experience, the experience of falling asleep, also, has as one of its characteristics that *we can return to it*; that is, that when we are in and with the world in our experience in a non-reflexive way, we are

3. Quoted in L.R. Müller, *Ueber den Schlaf* (Berlin: Springer, 1940), p. 38.
4. By "sleep" we understand the deep, dreamless sleep. In how far such a sleep occurs is not in question here because we may take it, at any rate, as a limit. Dream-consciousness to which a special chapter should be devoted is excluded from our further analysis.

able to turn back on it later in reflection in order to consider it.[5] And although it is true that the original experience itself, cut off from its vital context, then appears to us as an isolated and static *image*, we are nonetheless able to recognize in this image the original flowing being-one-with-the-things which characterizes our experience. Whereas we must call sleep, taken as pure life-phenomenon, *unreflexive*, we must call the unreflective experiencing and doing *pre-reflexive*. So although it is indeed pre-reflexive and therefore in the strict sense not reflexive, it nevertheless functions as the point of departure, as the turning-point for our spontaneous reflection, and as essentially permeated with the *possibility* for such a reflection. The fact that the experience can be reached by spontaneous reflection, and that there is a possibility of taking both back again in theoretical reflection offers the possibility of a phenomenological analysis of our falling asleep.

The question of falling asleep is asked here as a *psychological* question. That is to say we consider falling asleep as a specific relation between person and world. In examining the essential characteristics of this relationship we shall not separate "mind" and "body," and we certainly do not consider them as two substances, simply because they do not manifest themselves to us in this way originally. We conceive of mind and body as two aspects under which we can grasp the person in his relation to the world. It is this person who falls asleep, not just his mind or his body. Connected with this point of view is the fact that we are unable to formulate the psycho-physical problem of falling asleep: how does the mind work on the body, or the other way around, how does the body influence the mind. And in view of the fact that we refuse to reduce the human body to a thing, we can have no interest here in the "merely bodily" processes of sleeping. We shall lay aside here what physiology teaches us about sleep and falling asleep, that is all scienctific knowledge concerning certain *effects* which we can study in their causal lawfulness only by reducing the human happening to a *Geschehnis* without further qualification. Physiology considers sleep merely in a formal way, as a sign of events which occur in the organism. We wish to encounter falling asleep in its phenomenal content before any causal explanation and explain it as an intentional relation between person and world; hence as a game of "motives" and "decisions" which constitute a part of a meaningful history; as a limit-phenomenon which indicates the transition from being awake to being asleep, from knowing to living as such.

II. CONSCIOUSNESS OF FALLING ASLEEP

Normally falling asleep takes place in a few seconds. I lie down, relax, close my eyes, and suddenly I am "gone," departed for the land of dreams under the safe guidance of Morpheus. What takes place in these brief moments? We have a few

5. Cf. for the problematic concerning experiencing: P.Th. Hugenholtz, "Over het beleven en de belevingswereld," *Nederlandsch Tijdschrift voor Psychologie*, Nieuwe Reeks, 6(1950). Concerning these considerations we note, however, that we are unable to share Hugenholtz's ideas about the "autonomy" of experience. Cf. infra, sect. 4.

studies at our disposal which concern themselves especially with these moments which are difficult to grasp.[6] Trömner made a distinction between two phases in falling asleep. First *somnolence* or drowsiness, "the sinking away", this typical, pleasant state in which we still just about know what is happening. In somnolence we gradually lose consciousness and our state of mind is quiet and pleasant. One lies in bed and still plays a bit with his thoughts which then slowly evaporate and disappear. Then *dissociation* follows, at least according to Trömner. The conscious personality begins to disintegrate, loses its integration, conscious life loses its order. Thoughts emerge and disappear again suddenly; vague feelings, moods overpower us and overtake our conscious thought until finally sleep sets in.[7]

Angyal made an introspective study of falling asleep over a period of 20 months and then verified his data on the basis of reports by six other subjects.[8] He divides falling asleep into three phases. *First thought and volition are pushed aside by pure associations.* Our sensibility decreases and associated with this there is a loss of orientating unity of thought and will, which are the most conscious of all our functions. The determining tendencies, well-known from the Würzburg school, particularly suffer a loss. When the person who is falling asleep thinks about a problem he is no longer so intensely oriented toward finding a solution as is the person who is awake. On the contrary, he even forgets the task, wanders from his subject, loses his way, and gets completely lost in his sleep. This first phase leads straight over into sleep and dream, at least if the second and third phase do not insert themselves in between.

If the second phase of falling asleep occurs, it becomes the moment in which hypnagogic visions occur, the images characteristic of falling asleep which have been described in such a striking way by Leroy.[9] They are more or less optical in nature, although a number of them also originate from the bodily position of the sleeper and from stimuli affecting the sense of touch. Small, glimmering pictures appear and disappear, whimsical and unordered by fixed lines, continuously changing into a kaleidoscopic whole. Sometimes they occur on the basis of perseveration of representations, flowing from, or more or less joining, contents of consciousness present before the subject's falling asleep. They dance loosely around one another, bound by nothing except the laws of association. The characteristic

6. On the complementary phenomenon, namely waking-up, there is an interesting study by M. Grotjahn, "Über Selbstbeobachtungen beim Erwachen," *Zeitschrift für die gesammte Neurologie und Psychiatrie*, (139), 1932.

7. E. Trömner, "Vorgänge beim Einschlafen," *Jahrbuch für Psychologie und Neurologie*, 17 (1911).

8. A. Angyal, "Der Schlummerzustand," *Zeitschrift für Psychologie*, 103(1927). Cf. also: K. Leonhard, *Gesetze und Sinn des Träumens* (Stuttgart: G. Thieme, 1951²), pp. 99ff.; E. Claparède, "La question du sommeil," *Année psychologique*, 18(1912), pp. 456ff.; P. Schenk, "Ueber das Schlaferleben," *Monatschrift für Psychiatrie und Neurologie*, 72(1929). The changes in consciousness during falling asleep are characterized by Sartre as transition to a "captive consciousness". J.-P. Sartre, *L'imaginaire* (Paris: Gallimard, 1948), p. 55ff.

9. E.B. Leroy, *Les visions du demi-sommeil* (Paris: Boivin, 1933²). Cf. for an excellent summary of the data. H. Ey, *Études psychiatriques*, vol. I (Paris: Desclée de Brouwer, 1948), pp. 167ff.

trait of this second phase, in which there is no longer any question of thought, consists in *the dropping out of all consciousness of meaning*. The representations and events stand empty before us, they are just mere data, they do not have any meaning, and they do not appeal to us.

Finally, the third phase is characterized by the appearance of larger images and scenes which gradually lead us over to the dream. Together with these scenes there is *a return of self-consciousness*, which had previously disappeared. But this is a changed consciousness of self; it is already a dream-consciousness. We are present again all right, but in a way which is appropriate to the dream, in that dream way which we do not wish to determine any further here.

This totality of phenomena which Angyal evokes for us, refers to the same characteristic as we suggested in Trömner's descriptions: the *increasing dissociation* of consciousness characteristic of our falling asleep. That is why Angyal says: "Depersonalization, abrupt changes of location, dream-metamorphosis, and many other characteristics of the dream are natural, necessary consequences of the fact that our self-consciousness in the dream is built upon labile, swiftly changing, volatile foundations, on representations."[10] Hoche, also, speaks of the becoming more indistinct of contents of consciousness in a similar manner; they withdraw as it were from a center, thoughts slip away, unintended representations arise spontaneously; what we experience is meaningless. Hoche aptly speaks of "Leergang der Mühle"; just as the sails of a windmill just turn around and around, so the contents "spin" around in our consciousness.[11]

Yet these explanations by Trömner and Angyal are not completely satisfactory. No doubt we can verify their statements. And yet it seems to us that their investigations have missed the heart of the matter. The expression "dissociation" seems to be incorrect; for there is not really a question of a "falling apart" of our consciousness into component parts, such as "representations" or "contents" of another sort. When Bizette says that in falling asleep we are overpowered by the game of representations,[12] we cannot refrain from asking some questions about the nature of the *representing acts*. The time is past when man's psyche could be conceived of as a scene in which the events are governed by the almighty laws of association. Thus it is not so much the data described by Trömner and Angyal that we take to be incorrect, but rather their theoretical background. However, this does not mean that we can limit ourselves to just *re-interpreting* these data. The theoretical background and view have already *constituted* these data in a determinate way. And further, is it not also true that the method of investigation has contributed its share to the alienation of the real event we experience in the description? For while we are falling asleep we do not *experience* a falling apart of ourselves, no more than we experience an emerging of separate "representations." Although the term "dissociation" certainly refers to a phenomenon all of us know from our own experience of

10. Angyal, *op. cit.*, pp. 97-98.

11. A. Hoche, *Das träumende Ich* (Jena: G. Fischer, 1926).

12. A. Bizette, "Remarques sur les phases du présommeil," *Journal de psychologie normale et pathologique*, 28(1931).

falling asleep, in its spatializing connotation the term suggests a conception of the event which is certainly not adequate. We must guard against succumbing to the seduction of language which can too quickly force a theoretical construct on us.

Angyal always awakened himself after determinate intervals during his process of falling asleep. This involved a psychic awakening, an "attitude" which woke him after a certain course of time, so that he could write down what had happened. This method allegedly had the advantage that the slumber images did not become scattered.[13] The question, however, is whether they did not, indeed, change because of the interfering and fixating attention. Is it not true that the argument used in the battle against introspection, namely that fixation of our attention on the inner event changes this event, weighs even more heavily here and thus that the "data" are falsified?

Introspection discovers in our falling asleep contents of consciousness which stand before us as "meaningless." But when we turn back toward our own falling asleep, we must say that this so-called meaninglessness is not *meaninglessness* at all, but precisely a typical meaning structure which exactly characterizes our consciousness of falling asleep. There is no question of a "dropping out of our significative consciousness", but rather of a typical fading, a *change* in meaning. This is the "crude phenomenon" whose structure is to be uncovered in the analysis to follow. This central phenomenon has a double aspect and must be approached from both these perspectives. On the one hand, falling asleep is an *act-history*; and on the other it is a *change in the meaning of the world* which is correlative to this history.

But if it is already impossible to recover an act in its actual performance by a new act, how will we be able to recover a dying act and its very dying by means of a living and waking act? Is it not true that the real history of the acts which occur while we are falling asleep will forever remain a mystery for our consciousness? This is certainly not the case. For we are able to say a great deal about our waking acts although they, too, escape us in their actual execution. *The quality of the act can be discovered in and from the mode of givenness of its correlate*, in the full *noema* in which it lies "materialized" as no-longer-act, "materialized" in the *meaning* of its own correlate. Its being-an-act and its originating from the "center" of the person as act-origin can only be experienced in the experience of "*I-in-my-origin*".[14] It is not I taken as field of consciousness filled with contents, but "I-in-my-origin" who falls asleep; and this is connected with a qualitative change of the total-correlate of my acts which we may circumscribe as "my world". *My world which falls asleep supplies us with the guiding-clue for our analysis of the acts involved in falling asleep.* Thus in order to understand falling asleep we must examine the world of falling asleep.

But do these last reflections not make us return once again to Trömner and Angyal who already described this world of dreaming-away? Yes and no. They conceived of this world as a scene of separate, almost independent phenomena con-

13. Angyal, *op. cit.*, pp. 67.

14. Cf. the development of this concept in S. Strasser, *Het zielsbegrip in de metaphysische en de empirische psychologie* (Leuven: Nauwelaarts, 1950), pp. 57ff.

catenated and connected by the "gentle force" of association (Hume). One then calls these phenomena "representations" and gives them an independent existence of their own consciousness. True, they manifest themselves to consciousness, but they remain autonomous. We, however, wish to look through this world toward the "marrow" of the person whose world this is. We shall find the person who falls asleep by way of his world which falls asleep. We will find material for this analysis in our own experience and in the description which other people have given of this "world-falling-asleep." These descriptions will serve the purpose of evoking the phenomena so that we will be able to consider them.

That it makes sense to speak of a world which falls asleep has already been witnessed by Marcel Proust when he says: "I would fall asleep, and often I would be awake again for short snatches only, just long enough to hear the regular creaking of the wainscot, or to open my eyes to settle the shifting kaleidoscope of the darkness, to savor, in an instantaneous flash of perception, the *sleep which lay heavy upon the furniture, the room, the whole surroundings of which I formed but an insignificant part* and whose unconsciousness I should very soon return to share."[15] The home falls asleep, things become insensible and admit him in their sleepy sphere.

But if this is a suitable starting point for our investigation of falling asleep then we may take van den Berg's point of view in what is to follow. "The subject shows himself in things, at least if one dares to take these things in their original form, that is in the way they appear ..., or to use an expression by Straus, if one takes them in their scenic value."[16] We shall make this standpoint our own but add to it that the scene which is falling asleep refers to the acts of the person who is falling asleep and which are ungraspable in their actual execution.

C. Schneider who was obsessed by an alleged relationship between falling asleep and the schizophrenic experience has tried to perform an analysis of the act of falling asleep without using the world which falls asleep explicitly as his guiding-clue. We wish to follow this analysis here in that it is very useful as a preliminary description of the history of the act.[17]

Schneider says that in falling asleep the whole of our experience is changed in the direction of volatility, "impenetrability," and loss of order. The constancy, characteristic of the waking appearances gets lost, the person loses himself in the stream of experiences so that the experience itself has only the character of a "mere appearing." Or, to use an expression from Mayer-Grosz and Beringer, the experiences withdraw from the grasp of the intention. With this loss of mental activity there arises the unsharpness, "*Verschwommenheit*" of what is experienced. The

15. Marcel Proust, *Swann's Way*, trans. C.K. Scott Moncrieff (New York: The Modern Library, 1956), p. 4. Italics are ours. We may place "the psychical night" as H. Beaunis has called drowsiness, opposite this falling-asleep of the world; cf. "La nuit psychique," *Congrès international de psychologie de Rome*, 1905, p. 396.

16. J.H. van den Berg, "Menselijk lichaam, menselijke beweging," in *Nederlandsch Tijdschrift voor Psychologie*, Nieuwe Reeks, 5(1950), p. 296.

17. C. Schneider, *Die Psychologie der Schizophrenen* (Leipzig: Barth, 1930), pp. 12-20, p. 76, pp. 110ff.

emotions become weary so that one can speak of a certain affectlessness. It is as if nothing any longer interests the one who is falling asleep. All direction toward what is in the future and each glance toward the past loses interest. The ordering of our experience begins to show lacks; true the one who is falling asleep is still oriented in time, but he lives in the "now", or better in timelessness. Connected with this there arise "holes" in what is experienced; little brooks separate from the continuing stream of experiences which no longer blend with one another continuously, but acquire a certain autonomy and sometimes even overlap. Essential and accidental aspects lose their specific value and begin to mix. The experiences escape from the one who is falling asleep, they intertwine, and go their own ways. The need to express oneself diminishes; the world loses its value, one sinks into himself and no longer has any tendency toward an exchange with the surrounding world. Experience becomes "vast" and recedes to an ego-distance; events are still "taken in", but one's own activity grows dim.

What distinguishes these analyses from those described by Trömner and Angyal is not so much the fact that new content-like data are found here, but the fact that the event is seen in function of the I as the performer of the acts. It is the person who gives up his world, who withdraws, and therewith gives up the ordering of the experienced world in space and time, past and future, I-here and the rest over-there. While we are awake each change of intentionality is connected with a scission (*caesuur*) in the experience; although it is true that the stream of consciousness keeps flowing on, we can easily distinguish between one act and the next through the reflexive attention which grasps them. It is this intention that is recaptured while we are falling asleep.

Undoubtedly, Schneider's analysis, also, is subject to objections. Mayer-Grosz has justly corrected him on several points.[18] Indeed, it is not correct to speak of affectlessness. On the contrary, it is precisely the moods which begin to play the predominant role and carry the experience.[19] There is a change of direction in experience, and not so much a becoming-dim. There is no question of dissociation – by the way, Schneider would never have said this explicitly – as is clear from the occurrence of "intentional shells," that is of "free floating" acts which emptied of a determinate content are described as "interpunctions of a sentence" without words. Mayer-Grosz even speaks of "fields of forces of thought."

In this latter phenomenon we encounter a datum which Schneider undoubtedly must have had in mind, too, when he spoke of the taking-back of the "Eigentätig-

18. W. Mayer-Grosz, "Einschlafdenken und Symptome der Bewusstseinsstörung," *Archiv für Psychiatrie*, 78(1926); *Pathologie der Wahrnehmung*, II, in Bumke's *Handbuch der Geisteskrankheiten*, vol. I (Berlin: Springer, 1928), pp. 433-438; "Zur Struktur des Einschlaferlebens," *Archiv für Psychiatrie*, 86(1929).

19. H. Ey, (*op. cit.*, p. 172) writes the following: "The marginal phases of sleep are intensely affective. Through its contents the fascination of consciousness joins the world of images to the sollicitations of the instinct. This marvelous flowering-time produces a kind of Nirvana-state, the bewitching fiction of a foreshadowed dream from which one has not yet completely detached himself; the attraction of the world of images has us still in its spell and the one who falls asleep feels 'entranced' by the world of dreams."

keit' (the quality of self-acting). If one is to understand falling asleep this is a point of the greatest importance. The reflexive acts such as thinking, paying-attention-to, going-back-to, gradually empty themselves before dying altogether; the breaks found in the course of the acts disappear. The change in quality of our being-conscious which sets in with this does not imply an immediate disappearing of our being-directed-towards, but rather a letting-go, an abandoning emptying of the reflexive acts; for insofar as these acts can still be called acts they *must* have this '*Beziehungsbereitschaft*' (the readiness to engage in relations) of which Mayer-Grosz speaks and which, in addition, as we saw earlier, is described as the *idleness* of the windmill. All attempts to grasp and describe in words these de-activating acts in their actual execution (or in their withdrawal), are destined to fail. That is why the *image* evoked by the mill which keeps spinning idly, inadequate though it is, is still most suitable to represent the occurrence. The empty 'act shells' are like unspoken punctuations in a sentence, which are not filled up with act-correlates. For the withdrawal of the intentions precisely consists in this letting-go of the correlates which now, insofar as they are still contained in the experiential intentionality of the moods, receive a completely different appearance. The world itself divests itself of its breaks, loses its sharp contours and elaborateness.

Falling asleep is at the same time disintegration and integration. This is why Schneider as well as Mayer-Grosz are correct. The disintegration consists in the loss of the ordering-unity of consciousness which flows from the fact that the reflection withdraws, empties itself, turns back on itself and in so doing disappears; reflection thus becomes dissociated from the experience which thus obtains a certain autonomy. The contents of consciousness characteristic of our falling asleep acquire an independence and float along supported by the emotions. But this means that we can speak equally of an integration of the experiences. The experience sphere rounds itself off, becomes free and floating, not interrupted by the incising, reflexive acts. The remarkable thing here is that speaking symbolically reflection and pure experience, dissociated though they may be, "fall asleep" *simultaneously*. Experience reaches its complete autonomy and unreflexity only at the moment when the person is sleeping and thus no longer experiences anything. But before going into the relationship between experience and reflection we must first turn again to the *world* characteristic of our falling asleep since it is this world which must provide us with the guiding-clue for an anlysis of the acts.

III. THE STILLING AND DARKENING WORLD

There is a general conviction that falling asleep is closely connected with silence and darkness. How must we conceive of this relationship? It is to this question that we wish to turn our attention now. The current view is inclined toward the judgment that silence and darkness in *their negative and isolating function* are the causes or conditions of falling asleep; no reaction without a stimulus. In a work of the 17th century physician Van Beverwijck[20] who adopted Descartes' view and explanation we find the following: "In order to evoke sleep it is first necessary to remove everything that could arouse any of our senses: bright light, all kinds of sounds. The persons who wish to sleep must be kept completely quiet, put in a soft bed in which there are no mosquitoes and fleas to be found; they must close their eyes, concentrate their thoughts from the other senses on one of them, a process through which the Spirits, equally turned away, gradually become still." In our sleep influences from objects in the outer-world are prevented from penetrating the brain where they become awarenesses; on the other hand, the vital spirits who find themselves in the brain are now unable to move to our limbs in order to move them. According to Descartes[21] these two are the most important sleep-effects. We find the echo of this train of thought in Jaspers: sleep sets in the moment our psyche is isolated from the outer-world. "If, as is usual, tiredness does not yet overcome us, the main condition for sleep is a situation that reduces stimuli to a minimum: darkness, quiet, a peaceful mind, a relaxed position, absence of muscle tension. The complete exclusion of stimuli *induces sleep*."[22] The exclusion of movements and sensory processes, motionlessness and unfeelingsness induce sleep. At first sight it seems as if the heart of the matter is touched here. For if we wish to sleep the first thing we do is to stop moving, to lie down, to darken the room, to exclude all disturbing noises. We know how difficult it is to fall asleep when a toothache or any other pain tortures us, not to mention the fleas and mosquitoes Van Beverwijck spoke of. Thus sleep would indeed be an isolation of brain and "psyche". That is why Bremer obtained a typical deep-sleep oscillogram from the brain of a cat by isolating the brain; he cut the brain-stem after the origin of the third nerve and the brain which was then isolated, "slept".[23] But this can be done in a simpler way, also: by placing an animal in silence and darkness it can be brought to falling asleep. The exclusion of all possibility of motion leads to a similar effect. Says Buytendijk, movement can function only in connection with the sensory; but also conversely.[24] If we silence the sensory or the motor system, then the other of the two immediately loses its function also. And when we try to consult our own consciousness of sleep we, too, must come to the same conclusion, namely that regardless of what this consciousness may be, the outer-world has disappeared in any case. We even know that we have slept, because we experience having *returned* to the world.

Still further reflection gives us renewed insight: sleep is not caused by the falling away of the outer-world, nor by the cutting-off of all contact, but it is in

20. J. Van Beverwijck, *Schat der Ongesondheydt, ofte Genees-konste van de sieckten* (Dordrecht: Gorissz., 1651), p. 141.

21. R. Descartes, *Traité de l'homme. Oeuvres*, ed. Ch. Adan et P. Tannery, vol. XI (Paris: L. Cerf, 1919), p. 197. Cf. also in vol. IV (Paris, 1901), p. 192, in the letter "Au Marquis de Newcastle".

22. K. Jaspers, *Allgemeine Psychopathologie* (Berlin: Springer, 1948[5]), p. 196. Italics are ours.

23. F. Bremer, "Cerveau 'isolé' et physiologie du sommeil," *Comptes rendus de la Société de biologie*, 118(1935).

24. F.J.J. Buytendijk, "Le repos et le sommeil," *Traité de psychologie comparée* (Paris: Presses Universitaires de France, 1952).

falling asleep itself that the world loses its waking value and waking meaning. If we are to understand our falling asleep we cannot limit ourselves to the stimulus-response schema. The thesis we wish to defend is *that there is no univocal relation between the "exclusion of stimuli" and our falling asleep.* Darkness and silence can *motivate* our falling asleep, but they cannot cause it.

We must clearly distinguish between the pseudo-rest of the isolated animal and man's falling asleep. Buytendijk has explained that tightening of movements indeed gives the *possibility* of rest, but *is* for that reason not yet rest, and that a fixation of position can *go over* into rest only secondarily. Genuine rest distinguishes itself from motionlessness in that it is a *genuine function*, an exposing of oneself actively to the diffuse, enveloping situation, a surrendering, a removal of tensions, and an oriented *active taking of a position*, a settling down or relaxing. This rest, which we also know when we are awake, is a condition for falling asleep which comes about in a quiet atmosphere. In this way which is in agreement with the way Hess expresses it, sleep is no longer understood as a "*Funktionsdefizit*" (a functional deficiency) but as an act, as an *act of surrender.*[25] True, it is a most remarkable act, in that it leads to a refraining from all acts. We shall see how sleep conflicts are precisely founded in a transformation of this typical structure.

Falling asleep does not mean a becoming-cut-off from the world, but a quiet giving-up of the appeal the world directs to us. That is why the person falls asleep only *insofar as this appeal becomes silenced*, and he feels *at ease about it*. Each treatise on falling asleep cites the case of the mother who sleeps through any noises except the distant crying of her child. It is clear, James says in his witty way, that the baby-section of her acoustic sensibility is systematically awake.[26] There always remains a bit of interest in the things we cannot neglect without physical or moral danger. There are, for instance, people who wake up several times a night to determine whether the alarm clock which has to wake them up in the morning still works. We sleep only insofar as we are not with something, be it the baby or the alarm clock. Furthermore one never sleeps better than during a boring speech.

Thus it is certain that sleep and sensibility do not exclude one another.[27] It is

25. Buytendijk, *op. cit.*

26. W. James, *The Principles of Psychology*, vol. I (London: Longmans, 1890), p. 213. E. Cramaussel has collected a treasure of data concerning the sleep of the young child: "Le sommeil d'un petit enfant," *Archives de psychologie*, 10-11(1911) and 12(1912). Just as the mother while sleeping does not forget her child so the child while asleep is still "interested" in certain things. That is why the author says (1912, p. 183): "... that which gives evidence of an inner organization which is carried on, of a central work which forms and transforms itself, is the extraordinary importance which certain weak but suggestive excitations receive which do not wake the child up, but which *are of interest* to it in sleep as well as awake: the barely perceptible noise of a toy which it likes, the subdued voices of his brothers and sisters playing in the yard, the water one lets drip from a sponge. Whereas he remains closed in regard to excitations which otherwise are quite strong, he is always ready to prick up his ear for these and some of them leave a long echo in the child." On the contrary there is the fact that the child instead of waking up falls into a deeper sleep in the event of some disturbing noises, such as talking near his bed.

27. Cf. Claparède, *op. cit.*, p. 434 and also by the same author: "Le sommeil et la veille," *Jour-*

wrong to make of sleep a mere physiological problem. The problematic here is primarily of a psychological nature. When Ziehen said that "probably the essential thing in the coming about of sleep is found in the exclusion of outer stimuli ... and the weariness of the cells of the cortex,"[28] this was not only a hasty hypothesis, but in addition a pertinently wrong one. No one has taken this theory of isolation to the absurd more than Gorter. According to this author the efficient cause of sleep is the setting of the sun. For many functions of living matter are connected with light and diminish when light begins to disappear or even cease themselves. If man's sleep does not so obviously depend upon the setting of the sun, this is due to the fact that "man conducts the struggle for life with artificial light."[29] This is a remarkable inversion of the relationships. We do not go to sleep because the light is turned off, but we turn off the light in order to be able to sleep. The latter would be impossible if we would necessarily be awake while in light.

There is a relationship between falling asleep on the one hand and silence and darkness on the other; but this relation is not a causal or an effective one. There is an *essential relationship* here which we wish to make explicit by examining the meaning of silence and darkness.

The silence we need in order to sleep is not merely the *absence of noise*, but the *meaningless, stilling* silence. The ticking of the clock, a speech which is boring, the creaking of the bed, noise made by streetcars and cars, and even a lively conversation around us do not keep us from falling asleep provided they are meaningless and worthless; on the other hand, however, a low soft conversation in the room next door, or an alarming drip from a leaking roof, the irregular breathing of your wife are enough to deprive you of all sleep. This is because they address themselves to us and we have an answer to their appeal. And so it is not noise or absence of noise in a physical sense which maintains a relationship with falling asleep, but *silence*. "When I am awakened by a voice at the bottom of the stairway in the morning," van den Berg says, "then my answer 'sounds' where it is supposed to be heard: that is at the bottom of the stairway and eventually it is heard there so exclusively that it becomes exceptionally quiet around me, so quiet that I quietly sleep further."[30] I wake up insofar as I have to answer down *there*; but as I surround myself *here* with silence, I can continue to sleep. When one retires from a busy meeting in the

---

nal de psychologie normale et pathologique, 26(1929), p. 449 where the thesis is defended that "sleep is thus always in a certain sense partial." K. Landauer comes to the same conclusion: "Handlungen des Schlafenden," *Zeitschrift für die gesammte Neurologie und Psychiatrie*, 39 (1918), p. 333: "There are for each observer visible, meaningful activities of the sleeper." What follows this statement, however, shows lack of phenomenological insight: "The sleeper is not absolutely 'stupefied': he is able to act logically and with energy."

28. Th. Ziehen, *Leitfaden der physiologischen Psychologie* (Jena: G. Fischer, 1914[10]), pp. 397-398.

29. A. Gorter, "De oorzaak van den slaap," *Verslag der Koninklijke Akademie der Wetenschappen* (Amsterdam, Wis.- en Natuurkundige Afdeling), XII/1, (1903), p. 151.

30. J.H. van den Berg, "Het gesprek en de bijzondere aard van het pastorale gesprek," *Theologie en Practijk*, Nov. Dec., 1950, p. 164. Concerning silence see *Over zwijgen en verzwijgen* (Utrecht: Kemink, 1949), by the same author.

next room, it can be infinitely quiet in the new solitude; the conversation *remains* in the conference room – although I can still hear it –, but I am alone here with the silence. But if I had never attended a meeting in that conference room, and so had not had the actual experience of leaving it, then my concentration now in my own room would be lost in the hum coming to me from the next room; there would be no silence in my room, and thus I would be unable to work. What silence is, is determined in the meaning history of the situation which becomes constituted as intentional relation between person and world. The absence of noise is obviously a very strong motivation for the encounter of silence, but it is not a necessary condition.

But how can we now qualify the relation between silence and falling asleep. One of the characteristics of the world which falls asleep is that it *becomes still and wraps itself in silence*. When one speaks of "loss of contact with the outside world" we must understand it, as far as noise is concerned, not as an interruption of the flow of sound stimuli in a physiological sense, but as the stilling appeal of this world. The world becomes still and dozes off insofar as it no longer *addresses itself* to us. "For an animal darkness and silence are merely a lack of irritation," Buytendijk says. Man needs "to experience darkness and silence in their positive qualities"[31] in order to rest and sleep. Silence *is* something, namely that silent, enveloping something which we do not have to answer because it does not say or ask anything. And yet here, too, an important specification is still missing. It is not silence *in its positivity* which motivates our falling asleep, but the *stilling* silence: the stilling which we may circumscribe as the gradual becoming silent of the conversation. Conversation is to be taken here in its general sense as *dialogue between me and my world*. Thus the world does not get lost while I am falling asleep, but it merely becomes silent with me and with this common becoming-still both of us fall asleep. We cannot sleep in isolation, *outside* the world, but we can do so in a world that has become still and which, while we are sleeping, keeps us always safely protected.

Taken in its *negativity* darkness, too, is unable to motivate our falling asleep; nor can it do so taken in its *positivity*. What has already been said in connection with silence need not be repeated here. Neither the "absence of light" nor darkness as such correlates with falling asleep, but the *becoming-dark*. Here, too, that which maintains a relation with falling asleep is the meaning and signification, not the electromagnetic radiation. On a warm summer day one can sleep well on a brightly lit moor, whereas on the other hand a number of people precisely cannot sleep in "genuine" darkness. There are people who need a soft night light. For some people it suffices that the switch is close at hand so that light is continously present potentially. But these are cases where the person no longer dares to trust himself to darkness.

On what is the notion that it should be dark in order to be able to sleep founded? In order to understand this we must turn our attention for a short time to light. "On what is this pure joy founded, the experience which made Schopen-

31. Buytendijk, *op. cit.*

hauer write that light is the most enjoyable of all things...? It has its foundation in the fact that the eye in this case has the object in perception *in the way it is*; this expression means nothing other than the fact that the eye recognizes the object clearly and distinctly (objectively). That is to say, the object is no longer as it is at night when we realize with discomfort that our perception of things does not encompass them in their entirety ... that which has form must now manifest itself; it can no longer remain in hiddenness."[32]

Light makes manifest, says Heuss. It lays the perspicuous open and thus shows the unperspicuous things. It is only in the light that things can *manifest* themselves as they are. Light finds its fulfillment in the becoming clear of things, in the structuring and ordering of the world which surrounds us. It gives us a view of this world in its solidity and certainty, in its reliability, in its richness of colors and shades. Light makes manifest the world which is of interest to the eye and which prompts us to actions and tasks; it makes the world accessible and opens up the inviting distance. But above all it makes things *clear and distinct*. It is the sharpness of the bright world, its being cut into figures and backgrounds, the multiciplicity of points of view and particularities calling our attention which keeps us awake.

If the bright world is unfavorable for falling asleep it is because of its quality which flows forth from its brightness. A bright world means a *bright consciousness*, and particularly a *reflexive* consciousness. In a state of fatigue and in falling asleep the brightness-degree of consciousness becomes subject to a reduction, as we read in Bossard.[33] Is this just simply a metaphorical way of speaking? Or are we to hold the mysterious, inner relationship between light and reflexivity responsible for our being unable to sleep in light? Whatever the case may be, all illumination – taken in the double sense of the term – which is experienced as such, is detrimental to falling asleep. We can fall asleep as soon as evening begins to *fall*, taking away the brightness of the light, blurring it, and making things cease to manifest themselves. We fall asleep as soon as the world no longer attracts our attention. *This withdrawal of what manifests itself is the phenomenon proper of becoming sleepy.* That is why we sleep exceptionally well even in the light, on the condition that things no longer offer themselves as graspable to our attention. The only reason that night is the time most suitable for sleep, is to be found in the fact that when darkness sets in, the manifestation of things obviously diminishes and all sharp forms are taken back in darkness. While we are falling asleep, our consciousness becomes speechless and darkens in correlation with the silent world which withdraws in darkness.

This becoming speechless and this darkening must not be understood primarily as an event in the perceptual world, but rather as a change in the intentional relation between person and world, which can be motivated by the perceived events. *This* relation is essential for falling asleep. We wish to look at it once again in light of the knowledge that in some cases silence and darkness can precisely keep us awake.

32. E. Heuss, "Zur Metaphysik des Lichtes," *Neue psychologische Studien*, 6(1930), p. 261. Italics are ours.
33. R. Bossard, *Psychologie des Traumbewusstseins* (Zürich: Rascher, 1951), p. 77.

Just as some people need light in order to sleep, others need a ticking clock or some other noise. Darkness and silence motivate our falling asleep only insofar as they make something *fade out*. As soon as we encounter *darkness itself* and *silence itself* we can no longer sleep. When silence no longer means the gradual becoming-silent of the conversation, but the *concealment* of something that can be expressed at each moment, then silence becomes threatening and alarming. Then it appeals to us again and in such a way that we cannot escape it and so we lose all sleep. Disturbance over the threat concealed by silence, fear of this keeping-silent itself, keep us awake. As soon as there is no longer a stilling, but a *substantial* silence with which we even could start a *conversation*, then sleep becomes impossible for us. This holds true equally when the voice of silence is not threatening but encouraging, or even sweet, or perhaps sublime. Morpheus has lost all his power when we use silence for a conversation with what only speaks in silence, and darkness merely for an encounter with that which manifests itself only in darkness. For in this case we are again involved in a conversation with the world.

For this becoming dark, too, must be a dying-out. If, as according to the words of Hering, darkness places itself *between* us and things in order to cover them and fill space,[34] then we lose our grip on the world without having consented to it. Proust put it this way: "I regained sight and I was quite astonished to find around me a darkness, which was sweet and restful to my eyes, but perhaps even more to my mind, to which it appeared as a thing without cause, understandable, as a truly obscure thing."[35] Here darkness is truly an *obscure thing*, on which our eyes can continue to rest. Nothing is more frightening than this darkness which snatches things away from us. That is why Claudel says: "The night takes away our evidence, we no longer know where we are.... Our vision no longer has as its limit the visible, but the invisible as its homogeneous, immediate, indifferent, and compact prison."[36] Then we are lost in the night which has taken away our certainty in that it, as Rilke says, *gnaws* at our world:

Die schwarze Nacht sasz auf dem toten Tag,	The black night sat on top of the dead day
und Gott erschrak:	And God became frightened:
sein Blick ging lange in dem Dunkel irr;	His glance gone long astray in the darkness.
und als er trat aus Wolken und Gewirr,	And when He strode out of the clouds and confusion
fand er die Ferne nicht, nicht Flut noch Feld:	He found neither distance, nor flood nor land:
die schwarze Nacht frasz an der ganzen Welt.	The black night was gnawing at the whole world.[37]

In these lines the night becomes black and frightful. Here one *has to* sleep with light – that is if he can sleep at all, for even the lights are frightened:

34. E. Hering, In L. Hermann's *Handbuch der Physiologie*, III/1, (Leipzig: Barth, 1880), p. 573.
35. Proust, *op. cit.*, p. 3.
36. Quoted by G. Bachelard, *L'Eau et les rêves* (Paris: Corti, 1947), p. 140.
37. R.M. Rilke, *Das Buch der Bilder* (Leipzig: Insel Verlag, 1922), p. 55.

Die Lampen stammeln und wissen nicht:	The Lights stutter and do not know:
Lügen wir Licht?	Do we invent light?[38]

We can understand why Rilke says that on such a night God is the "only one who is awake and yet is not afraid."[39]

An encounter with *the* light is always sleep-stealing. The stepping back from the light need not motivate loss of certainty and anxiety; it can also mean a finding-again of intimacy and a feeling of safety in the encompassing, protecting darkness. What darkness takes away from us is the bright and sharp reflection, not the capacity to enter the life-world unconcerned and to wander in *rêverie*.

"When evening comes the life of the night begins in us. The lamp makes the dreams wait which are going to invade us, but the dreams come already into our clear thought. Our home is then at the borderline of two worlds," Bachelard[40] says; and the dreamer "lives locked up in himself, he becomes shutter, dark corner."[41] No one has known the night in this garment better than Baudelaire. For him the night became the genuine world, lit by an inner light, a home where alone one can really live. To the sun which has set he says: "Your memory shines in me as a monstrance!"[42] This is the moment in which alone one becomes himself, can return to himself. When the evening twilight falls he calls: "My soul, collect your thoughts at this grave moment!"[43] It is remarkable that only now does it become light for him. This peculiar light of the darkness which no longer lets us see clearly and sharply, but rather means fire, warmth, and the lustre of stars which warm our hearts. Baudelaire *adores* this night:

> Oh Night! Oh refreshing darkness! You are for me the signal of an inner feast, you are the release of anxiety! In the solitude of the plains, in the stony labyrinths of a capital, glittering of stars, explosion of lanterns, you are the fire of the artifice of the goddess Freedom.
>
> Twilight, how sweet and tender you are! The rose-colored rays which still linger on the horizon as the agony of day under the victorious suppression of night, the light from the candles which make blobs of opaque red on the last glories of the Occident, the heavy draperies drawn by an invisible hand from the depths of the Orient, imitate all the complicated feelings which struggle in the heart of man in the solemn hours of life.
>
> ... the flickering stars of gold and silver with which the sky is dotted, represent the fires of our fantasy which light up well only in the profound mourning of the Night."[44]

38. Rilke, *op. cit.*, p. 157.
39. Quoted by O.F. Bollnow, *Rilke* (Stuttgart: Kohlhammer, 1951), p. 54. Cf. chapter 8 of this work concerning the meaning of the night in Rilke's work.
40. G. Bachelard, *La terre et les rêveries du repos* (Paris: Corti, 1948), p. 114.
41. *Ibid.*, p. 98.
42. Ch. Baudelaire, *Oeuvres complètes* (Paris: La Girouette, 1948), vol. I, p. 45.
43. Baudelaire, *op. cit.*, p. 90.
44. *Ibid.*, p. 245.

IV. THE AUTOMATIZATION OF LIVED EXPERIENCE

If we have dwelled at quite some length on silence and darkness, it was not only in order to show that there is no univocal relation between the absence of light and sound and falling asleep. We have understood the function in which silence and darkness can motivate our falling asleep and also how this function is to be specified: *silence and darkness must make the world retire quietly*. If they do not do so and obtrude themselves and their possibilities of experience, as we saw happen for Rilke and Baudelaire, then they rather form obstacles to falling asleep. In this we have become acquainted with one of the forms of *insomnia*.

Insofar as this insomnia is founded in alarming silence and concealing darkness, further discussion is superfluous here, for one easily recognizes the anxious neurotic structures involved.

However, things are quite different as far as the positive *rêverie* is concerned. Does it make sense here to speak of insomnia? However grateful we may be for the works of Proust, Baudelaire, Rilke – and let us not forget Amiel – one cannot possibly doubt the neurotic character of their dream-like world. Baudelaire worshipped the night because of the figures which can only live there. Sartre witnesses of him that he did *not want* to sleep because he detested all surrender.[45] When in an unpublished preface to *Fleurs du Mal* (Flowers of Evil) we find Baudelaire conjure up sleep, he does so not because he wants really to sleep, but because he wishes to live in his dream-world in a conscious, but not reflexive experience. "To know nothing, to learn nothing, to feel nothing, to sleep and to sleep again and again, that is my only wish, an infamous and repulsive, but sincere wish."[46] When Baudelaire thinks he wants to sleep, it is merely so that he will not have to think, nor will, nor linger with things, so that he can keep drifting in the stream of his experiences. It is reflection that spoils experience, and it is this reflection which these neurotics flee. That is why Coleridge has sung the praise of Vishnu who floats about on an infinite ocean rocking on a lotus-leaf and each one million years wakes up for a few moments just to know he can sleep again for another million years.[47] What is the characteristic of reflection which so horrifies the dreamers? It is to be found in the fact that it cannot comprehend itself, that it alienates itself from the experience, and yet is presupposed for the possibility of letting oneself go in the stream of experiences. An animal does not know of *rêverie*; neither does the sleeper. In order to dream one must be awake, and it is this wakefulness of the mind which continually threatens to destroy the illusions of the dream, or at least to *fixate* them and so to deprive them of their life. This conflict is insoluble. Let us listen to Amiel's complaint:

> Laziness and contemplation! Sleep of the will, vacation of energy, indolence of being, how well I know you! To love, to dream, to feel, to learn, to comprehend, I can do all of that provided one relieves me of willing.

45. J.P. Sartre, *Baudelaire* (Paris: Éditions du Point du jour, 1947), p. 126.
46. Baudelaire, *op. cit.*, p. 472.
47. Quoted by W.J. Revers, *Die Psychologie der Langeweile* (Meisenheim am Glan: Hain, 1949), p. 21.

> That is my bent, my instinct, my defect, my sin. I have a kind of primitive horror of ambition, struggle, hatred, and of everything which disperses my soul making it dependent upon things and external goals.[48]

It is this static, shaped, fully structured outer world which evokes exasperation here because of its being illuminated. To be in this world means nothing except the fact of always having to be sharply conscious, to reflect, think, observe, establish, to remain distinctly and clearly aware of things, thus to know and to stay awake, to be out there, and yet to return to oneself and one's experience.

Can it still surprise us that anti-rationalists and protagonists of the philosophy of life place too much value on sleep? They conceive of sleep as a return to our "innermost self." The dream "that is that state in which you naturally find yourself again as you are, from the moment you abandon yourself, from the moment you neglect to concentrate on one and only one point, from the moment you cease to will," says Bergson.[49] The sleeper is quiet, "because he is blissful and satisfied, that is, because his soul has returned to itself from the harrassments of life, because it has returned home to its own satisfaction and no longer has to deal with strangers, because it has collected itself from distraction and in this collectedness is like the deep sea which shows a mirror-like surface as long as the storms are silent."[50]

Sleep is a point of rest; we cannot possibly doubt this. But is it true that sleep is genuine life? The destiny of man? Even Vishnu must wake up in order to *know* how delightful it is to sleep. Klages' thesis concerning the "unconsciousness of experience as such" is incorrect. In "pure" experience reflection is reduced, but it does not disappear. Sleep and especially dreams are glorified mainly because reflection is reduced here to a minimum without disappearing completely. And it is precisely this minimum which still makes an experience of our experiencing, a knowing of the blessedness of our not-knowing possible. The mind, this "adversary" of life, is for Klages almost an usurper, a *parasite*. "In the activity of waking there dwells the continuous inclination toward the passivity of not-waking and a letting oneself float without any steering."[51] That is why waking is strictly speaking not just being-awake, but a continually repeated being *awakened,* a being torn away, "where the soul believes itself to be torn, as it were, from the protecting arms of the mother toward the inexorable light and in the grip of a mysterious nostalgia, ominously becomes aware of the hidden treasures of her nightly life."[52]

48. H.F. Amiel, *Fragments d'un journal intime*, I, (Geneva: Georg and Company, 1919), p. 168. On p. 81 Amiel speaks of "our consciousness which immerses itself in the shade in order to take a rest from its thought."

49. H. Bergson, "La rêve," *L'Énergie spirituelle* (Paris: Presses Universitaires de France, 1949[52]), pp. 103-104.

50. J.E. Erdmann, *Psychologische Briefe* (Leipzig: Geibel, 1882), p. 116.

51. L. Klages, *Der Geist als Widersacher der Seele*, III/i (Leipzig: Barth, 1932), p. 807. Cf. M. Palagyi, *Naturphilosophische Vorlesungen* (Leipzig: Barth, 1924[2]), p. 218.

52. L. Klages, *Mensch und Erde* (Jena: Diederichs, 1937[5]), p. 52. See also what Rilke (*op. cit.*, p. 33) says about the man who is awakened by light:

"People are fearfully disfigured by the light
that drips from their countenances,
and if at night they have foregathered,

Here a mysticism of sleeping begins to develop which must make all wakeful thinkers suspicious. This making of our experience into an absolute which, independent of all reflexivity, would be pure experience and which would have to be reached through a willing not-to-will, points to a conflict and therefore to insomnia.

The unreflexive autonomy of lived experience is a theoretical construct which rests on a misunderstanding. Klages' conception of "being-conscious" as a being-reflexively-*knowing* and the consequence which follows namely that experience therefore is not conscious and must be rediscovered and reevaluated in its "unconscious autonomy," fails to appreciate reflexivity as an original human phenomenon. The basic thesis of phenomenology, namely that all consciousness is consciousness-of, holds true for experience also. Our lived experiences, too, are intentional phenomena, albeit pre-reflexive and thus *acts* in the broadest sense of the term. There is no ground here for identifying intentionality with *reflexive* intentionality. If the description of intentionality as "directedness-toward" seems to suggest this identification, then it is merely a consequence of the seduction of our language. If one wishes to characterize lived experience as *resonance*, he should not forget that in this resonance, too, we find a conversation, a dialogue with, and a being related to the *world* which is defined here as the entire correlate of all our intentional acts.

These reflections which strictly speaking belong to the general doctrine of intentionality may serve the purpose here of letting us recognize that the theoretical automatization of lived experience, as it has been defended by Klages, is a construction. And the latter is again of importance if we wish to evaluate the "factual" automatization of experience correctly.

Falling asleep is to abandon all explanation, even the unreflective conversation. The fact that in falling asleep it is primarily the reflexive acts which cease, does not mean that falling asleep is a return to the innermost self and its genuine world, the world of immediate experience. All of us know from our own falling asleep that phase in which reflection stops and the sphere of immediate experience undergoes a certain "rounding" which permits us "to float comfortably along in the stream" and to experience spontaneous fantasies. However, this is a phase in which the stilling does not cease but continues in order to result finally in the unconsciousness of sleep. This "autarchy" of our experience is already known to us from our waking-resting which characterizes the relaxed final phase. It is never unreflective but always open to reflection and capable of being relived in reflection, and co-determined by this possibility in its essential structure.[53]

---

you look on a wavering world
all heaped together."

(English: *Translations from the Poetry of Rainer Maria Rilke*, trans. M.D. Herter Norton (New York: W.W. Norton, 1938), p. 63).

53. Cf. the characterization by H. Ey, (*op. cit.*, p. 168f.), which we feel compelled to quote *in extenso*: "One could not better express that what characterizes 'hypnagogic consciousness' is the fact that it constitutes a metamorphosis of consciousness which becomes 'consciousness-which-makes-vivid' (*conscience imageante*). What circulates in its movement is not an idea; what organizes it is not an effort; what animates it is not a will. It is concrete, passive, and automatic

The fact that we keep dwelling in the world of experience prevents our falling asleep. Our invoking and encountering darkness and silence and our entering their emotional world means to start a conversation which keeps us awake, unless we again abandon even this mysterious world.

V. THE ABANDONMENT OF ACTIVITY

In falling asleep we must abandon all activity in the pregnant sense as "the accomplishment of acts". The sleep conflict is founded on our inability or unwillingness to abandon activity. We see this most sharply in the reflexive sleep-conflict which occurs during those hours in which we *want* to sleep and for that reason precisely are unable to fall asleep. The night slips by while the insomniac tosses and turns in his bed, exerts himself – but does not sleep. Sigwart once said that to understand the will we must start with an analysis of the will where we are most sharply aware of our willing as a determinate act,[54] – and this seems certainly to be the case in our falling asleep when it does not succeed, where the will experiences itself as powerless. Ach says that in the energetic act of the will we find an "activity" and an intentional object.[55] Without a doubt in the conflict of falling asleep we encounter this "activity" again, namely as pure effort – but where and what is the intentional object? Sleep is nothing, at least not something which our will can reach, because one must be awake in order to will. Here the solution to the problem conceals itself: willing *is* being awake; willing is keeping oneself consciously busy with something toward which one directs himself in all clearness and distinctness in order to materialize it. In other words it is obvious that willing to fall asleep is destined to fail precisely because it keeps us awake. The art of falling asleep consists in not-to-will. But to what peculiar problematic does this lead us? "The will must become suggestion," Jaspers says, "an agreeing and an expecting; it must become passive in its activity. It must not will to compel, but to surrender."[56]

---

as if it were stripped of that which in a wakeful state reflection, constraint, and perspective introduce to its structure. It traps itself and flows back toward the spectacle which it produces in and through its own movement. Completely fascinated by the imagery it brings forth and in the magical coalescence undoubled from what it is and what it does, it breaks its totality into fragments which are alien to its spontaneity. In this way nothing is born in this consciousness which does not shine as an image. But this overthrow of the world which stays at a distance from me, for which my being, seen and felt on the screen of the imaginary substitutes itself, *does not go so far as to abolish all consciousness of the game to which I surrender myself.* The miracle which takes place remains fragile *and within my reach. I experience that the marvelous event comes forth from me*, and if I give way to it then this is with the vague feeling that it takes the form of my desire to dream." (Italics are ours.)

54. Chr. Sigwart, "Der Begriff des Wollens und sein Verhältnis zum Begriff der Ursache," *Kleine Schriften*, 2e Reihe, (Freiburg im Breisgau: Mohr, 1889[2]), p. 118.

55. N. Ach, *Ueber den Willensakt und das Temperament* (Leipzig: Quelle und Meyer, 1910), pp. 240ff.

56. Jaspers, *op. cit.*, p. 197.

It is obviously meaningless to ask whether our will wills something or does not will something. Our willing cannot become substantialized into *the* will; it is just a title for a class of acts, for a determinate, specific way in which the person can have an intentional relation to his world. Now, willing-to-fall-asleep evokes something, directs itself to a world in order to start a conversation with it; but if one wants to fall asleep this conversation must cease and the world must retire. All acts have the world as their correlate and that is why falling asleep can never be an act, and that any accomplishing of acts is incompatible with falling asleep.

Woe to the man who thinks he can live only in activity. He must lapse into fear of falling asleep since his real possibilities for action are taken away from him by sleep. He has no confidence in the anonymous willingness of the body to wake up when it is necessary to take a position; he is afraid he will lose the grasp, certainty, and security of the world of those who are awake. "Do you know the fear of the one who falls asleep?" Nietzsche asks. "He is frightened down to his toes because the ground gives way under his feet and the dream begins."[57] Everyone has had the experience of awakening with a start when he was almost asleep because he thought he was falling. Why is it that we precisely *fall* asleep? For once we are in the land of dreams we have ample possibilities at our disposal to rise, fly, and float. But no one rises when he is falling asleep.[58] The ground gives way, the world retires and he who falls asleep slides back into unconsciousness.[59] When this sliding is experienced as falling it is anxiety over losing the certain world and the safe being-awake.

But let us return to *not-being-able* to fall asleep. Anyone who is occupied with something does not fall asleep. Anyone who is ahead of himself, moves toward the future, and thus is active, stays awake. Minkowski has shown how activity is an unfolding which implies a determinate temporal structure. Well, anytime the moment is surpassed, insomnia governs. Falling asleep requires a *suspension* of time, a no longer being addressed by what comes and what has been, a being taken up in this moment which then loses its moment-character and becomes timeless. If we do fall asleep we experience the time between being awake and sleeping as a timeless time, a suspended time. How can we make this experience explicit? Only by evoking it are we able to make it be present. In so doing we take the example of Zarathustra who one hot day at noontide lies down beside a tree and falls asleep. He even for-

57. F. Nietzsche, *Also sprach Zarathustra, Werke*, Band 7 (Leipzig: Kroner, 1964), p. 215.
58. There is a passage in R.M. Rilke (*Die Aufzeichnungen des Malte Laurids Brigge, Ausgewählte Werke*, vol. II (Leipzig: Insel-Verlag, 1950), p. 128) which seems to contradict this. He writes that Abelone "at that time was able to fall asleep without growing heavy. The expression 'falling asleep' is by no means appropriate to this season of her young maidenhood. Sleep was something that ascended with you, and from time to time your eyes were open and you lay on a new surface, not yet by any means the highest. And then you were up before dawn; even in winter, when the others came in sleepy and late to the late breakfast." This ecstatic puberty experience is certainly not a falling asleep, but a genuine ekstasis and rapture. The body loses its weight not through a weakening of the grasp, but through "sublimation." Here the issue is more about a living passing-away than about falling asleep. (English: *The Journal of my other Self*, trans. M.D. Herter Norton and John Linton (New York: Norton and Company, 1930), p. 140.)
59. Claparède strikingly speaks of a "slipping into sleep" (*glissement dans le sommeil*), "Le sommeil et la veille," p. 444.

gets his thirst in order to sleep at the perfect hour of noon. And in falling asleep he speaks thus to his heart:

> Soft! Soft! Has the world not just become perfect?
> What has happened to me?
> As a delicate breeze, unseen, dances upon the smooth sea, light
> Light as a feather: thus – does sleep dance upon me.
> My eyes it does not close, my soul it leaves awake.
> It is light! truly! light as a feather.[60]

"His eyes sleep does not close"; and yet he falls asleep. Zarathustra no longer knows whether he is awake or asleep. The world becomes round and perfect; but this means, also, that it closes itself and releases Zarathustra from his obligation to take a position: he may float around until the stream has lulled him asleep: the streaming sleep, light as a feather. We see how the body which relaxes in falling asleep gives up its position and man his relation to the world. The body becomes light and disappears because its grasp relaxes. This is the happy time, the solemn hour when even the shepherd no longer plays his flute. Therefore, do not sing, Zarathustra says, but be quiet. Be quiet and do not even whisper.

> ... soft! old noontide sleeps, it moves its mouth; has it not just drunk a drop of happiness – an ancient brown drop of golden happiness, of golden wine? Something glides across it, its happiness laughs. Thus – does a god laugh. Soft!

The old noontide, a drop of happiness, old and golden brown as wine. It is something old and familiar which is evoked here; that which has become so familiar and obvious that it no longer calls for any attention but carries us merely in its emotional character. He who wishes to fall asleep should turn toward the past; not toward the guilty and reproaching past; not toward that which lies behind him as something which, because it is unfinished, calls him back; but toward the familiar land of the child who did not know of cares and concern. The relation to the past may not be a relation to past *time*; one may evoke only the hazy and dreamed-away past in a mood which motivates a suspension of time.

> What has happened to me? Listen! Has time flown away?
> Do I not fall? Have I not fallen – listen – into the well of eternity?[61]

But then Zarathustra has fallen asleep. In an inimitable way Nietzsche describes his struggle when he realizes that he is asleep and must get up and yet still wants to keep sleeping.

60. Nietzsche, *op. cit.*, p. 401 (English: p. 287). See for the meaning of noontide in Nietzsche: O.F. Bollnow, *Das Wesen der Stimmungen* (Frankfurt: Klostermann, 1943^2), pp. 195ff. Amiel compares the noontide-mood explicitly with the night (*op. cit.*, p. 167): "Noontide; profound peace, silence of the mountains notwithstanding a full house and a village close-by. One hears only the sound of the fly which hums. This calmness is striking. The middle of the day resembles the middle of the night. Life appears suspended although it is most intense."
61. Nietzsche, *op. cit.*, p. 402 (English: p. 288).

> Up! (he said to himself) up, sleeper! You noontide sleeper!
> Very well, come on, old legs! It is time and past time,
> You have still a good way to go. ....
> (But then he fell asleep again, and his soul contradicted him
> and resisted and again lay down). "Let me alone!
> Soft! Has the world not just become perfect?
> Oh perfect as a round golden ball."[62]

This is the sound struggle of the noontide sleeper with himself. But one of two things must happen: either the world must keep silent and become still, and time must refrain from structuring the future – and then he will sleep; or time will call him back to the work that waits – and he will wake up.

When one is returning from a party at dawn and says: there is too much light already, I can no longer sleep!, this does not mean that light prevents our falling asleep but merely that we are responding to the call of light because we have already started the new day. He who is busy again already, or is still busy and thus gets ahead of himself in his activity no longer falls asleep. That is why desire – which Minkowski calls *le désir* – keeps us awake, too. In desire we have even gotten more ahead of ourselves, we are already more keenly, and more intensely, occupied with what is still to come. Any time the future addresses us when we begin to doze, we stay awake. Regardless of whether the time structure is that of activity or desire, or whether it is that of expectation (*attente*) in which we live time in an inverse direction (*en sense inverse*), it is in the anxious, paralyzing expectation of the coming threat that the future is awake.[63]

To fall asleep then means to disengage from what is coming, to let things take their own course and, unconcernedly, to take distance from them. During the storm Christ slept unconcerned when his small boat threatened to sink. Alexander the Great slept so deeply on the eve of the battle with Darius that he had to be called three times because the hour of the battle was close. Augustus was able to accomplish this, too, before the sea-battle with Pompey. The Emperor Otho was even able to snore the night before he committed suicide, after he had settled his affairs, divided his money, and sharpened his sword. And Montaigne was so amazed over these facts which he borrowed from Plutarch and Suetonius that he devoted one of his essays to sleeping.[64] Indeed, this is the provoking mystery of falling asleep, that we no longer bother about anything, or what amounts to the same, that the world retires and loses its appeal.

But how are we to force the world to retire if it refuses to do so? This brings us to the question of methods for falling asleep; this is a question which has been answered most completely by Jean Paul.[65] The author does not speak of veronal and similar poisons by means of which one *steals* from himself part of his sleep, but about the old and well-proven suggestive method. However he warns the reader in

62. *Ibid.*, p. 403, (English: pp. 288-289).
63. Cf. E. Minkowski, *Le temps vécu* (Paris: Presses Universitaires de France, 1933), pp. 79ff.
64. M. De Montaigne, *Essais*, I (Paris: Garnier, 1872), p. 288.
65. Jean Paul, "Die Kunst einzuschlafen," *Sämmtliche Werke*, vol. 52 (Berlin: Reimer, 1828).

advance that none of the means he recommends works, because a person exerts himself too much in using them. They have cost me enough sleep, he sighs.

The first method is counting, simply say the numbers one after the other. One must realize here that it is not important if he skips one of them once in a while: the tendency to be exact is disastrous here. The counting should be done as slowly as possible. The greatest possible "*Faulthierlangsamkeit*" (slowness of a sloth) is the most important part of the prescription in the methods for falling asleep. Then there are the notes, mainly the inner notes. There is no sweeter lullaby than the inner hearing of hearing itself. Furthermore, one can fall asleep by endlessly reciting short sentences: when the trees grow, when the sheep walk, when the clocks tick, etc. That Jean Paul is a good observer is clear from his statement that one must never let a conclusion follow these short sentences: no "then" may follow this "when." For in so doing one would construct a real sentence, a relationship with a coming effect; and any connection of sentences would be detrimental. Also, when it is a question of falling asleep *again*, one can put himself back in the interrupted dream, or internally look at something for a long time, stare into black darkness with closed eyes, turn back toward the past, jump swiftly from one subject to another, give his association free play, pay attention to the sounds of the body, tell himself a story, spell long and strange words, lift his fingers and let them fall down again, or finally evoke a scene, whatever it may be, in an endless repetition. As far as the latter is concerned we think of counting sheep or of the other excellent possibility which Jean Paul recommends: place yourself on a star and from a basket throw flowers into the universe until it is totally filled.

Thus the art of falling asleep comes down to the art of *boring* oneself, "an art which for assumed logical minds comes down to the illogical art of not to think."[66]

But Jean Paul certainly does not touch on the heart of the matter here. For in his analysis of boredom, Revers has shown that in cases of boredom time has become the object to which the rapt attention directs itself. And this tension is permanent because there is no solution possible in view of the purposelessness of the aspiration. Everything we are consciously aware of in boredom, is "old", it does not "*capture*" us, the future brings nothing new, but merely repeats the same boring old stuff. The time consciousness characteristic of the "*langen*" "*Weile*" (the "long" "while"; *Langeweile* is boredom) consists in the rationalization of the lost hope of still finding a goal in the future toward which one's aspiration could direct itself. Empty time is long. In boredom the waking I has lost its experiential contact with the world.[67] And although, according to our data, this indifference (*Gleich-*

66. *Ibid.*, p. 83.

67. Revers, *op. cit.*, pp. 60-61. See for the opposition between the *slow* form of temporalization of boredom and the rapid temporalization of sleeping L. Binswanger, *Grundformen und Erkenntnis menschlichen Daseins* (Zürich: Niehans, 1942), p. 474: "In contradistinction to *boredom* (*Langeweile*) one can determine it (although perhaps not as pastime (*Kurzweil*),) then certainly as a *short while* (*Kurzeweile*). Whereas there time appears to us as 'infinitely long', here it is 'infinitely short'; it seems as if between the moment of falling asleep and that of waking up, regardless of whether the sleep lasted five minutes or five hours, 'no time' at all has flown. This can mean only that the dreamless sleep ... is an extremely 'rapid' way of temporalization."

gültigkeit) in regard to the world would have to lead to falling asleep, it seems to us nonetheless that it is precisely this aspiration and need which will keep sleep away, as empty as they may be. The empty future, too, as future, keeps arousing tension and aspiration which, even though they have no object, still hang on to the world as bored world. One is unable to sleep because of boredom.

Is it then perhaps repetition itself, or monotony, rhythm which make us fall asleep? Klages believes that rhythm is responsible for this: "the experience of rhythm ... should, the more it gets the upper hand, release all tension and therefore among other things be capable of leading us to sleep."[68] Among other things, yes; but among these other things we will also have to reckon with jazz which is least of all soporific. Klages opts for rhythm because it belongs to life whereas a sharp metre belongs to the realm of the mind. But one cannot possibly maintain that rhythm is soporific! One can say that it draws us back from reflexivity into the world of moods, but this by no means guarantees sleep. No doubt rhythmical music, too, can motivate sleep; but then merely on the condition that it no longer addresses itself to us as rhythm but merely as a murmuring-along which fades away and does not ask for a response. The essentially soporific aspect of the lullaby, therefore, does not lie in the fact that it rocks us to sleep through rhythm, but rather in the fact that something is said to which we do not have to respond. That is why one can substitute the telling of an old and well-known story for the song; it evokes a sphere but no longer addresses itself in a way that requires one to take a position.

The lullaby is a *gratuitous story*. And yet it is not just something said or sung. In this the lullaby is precisely essentially distinguished from the monotonous and soporific noise of a machine. Monotony and repetition are soporific because of the continuously increasing emptying of the appealing character of what is repeated. What makes us fall asleep is not the constant monotony as such but the fact that in the repetition one becomes drained, and the fact that together with this the possibility of a conversation dies out or correlative with this, the fact that the accomplishing of acts becomes de-activated. The methods for falling asleep recommended by Jean Paul work mainly along these lines. The lullaby, on the other hand, is by no means monotonous. It is a *dialogue* (for it presupposes two partners) *in the form of a monologue*. I am addressed and I do not have to answer; this "address" does not call me to account for something and it does not aim at an explanation, but testifies that I am *sheltered and safe* under the protection of the one who watches over me. Whereas the lullaby (*slaap*lied) is soporific insofar as it is a gratuitous conversation, what makes it a cradle-*song* (slaap*lied*) can be found in the fact that it evokes a feeling of safety.

What we have noted earlier is confirmed here once again: falling asleep is not isolating oneself but a "drawing oneself in." We can do this safely only in a milieu characterized by contentment and the feeling of being secure. In order to be able to fall asleep I must know that I am safe; or perhaps better, I must unreflexively live the security of the situation in its enveloping and protecting character. I must be secure in regard to what obtrudes and threatens me in order to dare to relax and to

68. L. Klages, *Vom Wesen des Rhythmus* (Munich: Barth, 1955), p. 44.

sink away into the forgottenness of sleep. That is why we always find the characteristics of comfort and security in the sleep situation. The question of how these become materialized in each individual case, is something determined in the personal meaning-history. It may be that it is the strong yet soft arms of mother which guarantee the feeling of security;[69] perhaps it is merely her lullaby, or even her presence in the room or the house. But it can equally be the case that it is the hot-water-bottle at my feet which makes its comforting heat penetrate me, and in this way procures my access to the intimate and stilled sleep-world. It is possible that I must feel the hand of someone I love in order to know that I belong to a community which guarantees my security; however it may also be that I must precisely turn away from her and can sleep only with my back turned to her because only in this can I find my way to privacy and unconcern. My personal relation to the other and the community determines the way in which my falling-asleep-situation constitutes itself as a secure situation. But it is the essential characteristic of this feeling of security (which can be materialized in a thousand ways) that makes falling asleep possible for me. For the dis-activation of my activities, the abandonment of all interest, which is the central and essential characteristic of falling asleep, can be materialized only in this milieu. That this dis-activation indeed presupposes relaxation and rest and yet is still something other and more than this, is what we wish to examine now in connection with the part our body plays in falling asleep.

VI. THE BODY WHICH FALLS ASLEEP

Until now we have considered falling asleep mainly from the aspect of the mind. When we consider the body we will see that what has been said thus far can be confirmed, although it now looks slightly different.

In order to fall asleep the first condition is that we give up our *verticality* in one way or another. Vetter has expressed this nicely: not only the tension between our inner life and the outer world is abandoned but also that between above and below as is meaningfully indicated by the rest-position of the sleeper. The transition of the upright position toward the horizontality of the ground signifies the *abandonment*

69. Rilke testifies to this function of the mother who brings peace and security in an unsurpassable manner: "O empty night! O dim out-looking window! O carefully closed doors! Customs of immemorial standing, adopted, accepted, never quite understood. O silence in the stair-well, silence in the adjoining rooms, silence high up on the ceiling! O mother, O you only one, who put aside all this silence, once in my childhood. Who took it upon yourself, saying: 'Do not be afraid; it is I.' Who had the courage in the dead of night to be yourself the silence for the terror-stricken child, the child perishing with fear! You strike a light, and the noise is really you. And you hold the light before you and say: 'It is I; do not be afraid.' And you put it down, slowly, and there is no doubt: it is you; you are the light around these familiar, intimate things, that are there without afterthought, good, simple, unambiguous. And when something stirs in the wall, or a step is heard on the floor, you only smile, smile, transparent against the light background, on that fear-stricken face that looks searchingly at you, as if you were one, and under seal of secrecy with every muffled sound, in concert and agreement with it." (English: p. 71).

of the oppositions which characterize being awake.[70] Whoever goes to sleep lies down, comes to rest. This means that the watchful grasp of the body on the world relaxes. Straus has explained that the upright walk of man, his vertical *posture*, has a deep, anthropological meaning. Man's posture points upwards, away from the ground. His standing up which is already a first victory over gravity, has therefore the immediate expressive character of a pushing through toward freedom, toward height which has become the dimension of what is lofty and worthy of a human being. The direction downwards, collapsing, and falling on the contrary, thus becomes the expression of physical or moral defeat.[71] Anyone who thinks that this is no more than a romantization of a natural datum may recall the fear of falling asleep. Anyone who is afraid to *fall* asleep and who wakes up with a start from his falling-fantasm is really someone who cannot *resign* himself (*zich neerleggen*, "put himself down") to the fact that to sleep he has to give up his posture and freedom of action. All that the upright walk and posture mean to us, the opposition of person and world, standing up, reaching out to what is above, access to what is remote, the readiness-to-hand and graspableness of things around us, the surveyableness of the space in which we find ourselves, the choice of the place where we wish to stand – all of this is given up when we fall asleep.

Sleep is a *meditation of the body* which surrenders and relaxes. It abandons its role of being my grasp on the world: the body loses its "notion of task", albeit not completely as is the case in death. Three points are of importance: in this context: *decrease of sensibility, relaxation, and the "disintegration" of the body-schema.*

The first point, namely decrease of sensibility, is usually considered a physiological problem and in that sense is dealt with as a "raising of thresholds." That in dealing with this problem physiology by no means grasps it in its originality, was already shown in the third section of this essay. First of all there is no question here of cause and effect. Is the raising of the pain-threshold a consequence of falling asleep, or vice versa? In view of the fact that the issue here concerns not a causal, but an essential relation, the formulation of the problem is incorrect. Can sleep be thought of without decrease of sensibility? Even this question becomes an occasion for a pseudo-problem if one were to mean that "sensibility" is to be taken here as a property of the body conceived of as *thing*. Since the time that Straus recalled the original problematic of man's sensibility by re-qualifying the senses as communicative organs,[72] we have been able again to consider the sensibility problem in light of the relation between person and world. The body functions in this relation as a medium through which the person can adopt an attitude and enter into a relationship. In view of the fact that it is the relation-to, which along with the posture, is given up in falling asleep, it is obvious that our sensibility not only becomes useless, but even detrimental to falling asleep. For each sensitive relatedness to the world motivates an active or a re-active attitude, albeit perhaps merely as unspecified holding-oneself-ready-for. When we cease to speak we lose our audience.

70. A. Vetter, *Die Erlebnisbedeutung der Phantasie* (Stuttgart: Klett, 1950), p. 118.

71. E. Straus, "Die aufrechte Haltung", *Monatschrifte für Psychiatrie und Neurologie*, 117 (1949).

72. E. Straus, *Vom Sinn der Sinne* (Berlin: Springer, 1935).

In falling asleep our body loses its *incarnating function* in its double aspect of sensory and motor functions. Muscle tension diminishes along with the decrease of sensibility. According to Jacobson insomnia is always accompanied by residual tensions in the transverse muscles, that is to say in the musculature insofar as it is conducive to the relationship of the person to the world. Then we find here thus again what we have already observed earlier: falling asleep is abandoning this relationship, and for the body this means relaxation. Jacobson has founded a therapy based on this idea which is known as "progressive relaxation," a difficult and protracted process which aims at teaching man how he can relax by removing all remaining tonus.[73] The remarkable aspect of this relaxation is that the decrease in muscle tonus which in a certain respect is a negative process, rests on a *positive action*, namely the bodily equivalent of the feeling of resignation as Buytendijk expresses it.[74]

In view of the fact that we know that any activity is detrimental to falling asleep, it is useful to consider this remarkable activity somewhat more closely. Let us reflect on the moment in which we make ourselves ready for sleep in bed. Do we indeed *make* ourselves ready? *Do* we indeed *close* our eyes? During the day our eyes are open, at night they are closed. But is the transition from one state to the other an *activity*? The closing of our eyes is in a certain sense a sign of the approaching sleep, particularly when it occurs *unintentionally*; this is the judgment of Kleitman[75] and we can do nothing but subscribe to it. Obviously we can *close* our eyes, for instance at the sight of a horrible spectacle. But at night they fall closed, just as they "fall open" in the morning. The closing of the eyes when one goes to bed is something which constitutes a part of the complete situation into which the body enters. And the same holds for relaxation even though it is never complete. It is not correct to think that in going to bed we intentionally relax: we *become* relaxed in falling asleep. Our activity consists in a disposition which can be defined neither on the bodily nor on the mental level, namely the disposition of undergoing sleep. If this giving-up of all activity is, itself, again an activity, then it certainly is an activity of a special order.

The methods for falling asleep recommended by Jean Paul came down to making the meaning through which the world urges us to adopt an attitude and to take a position, turn pale. In falling asleep the last activity to be performed is that of evoking a world which falls asleep and stops appealing to us. It seems to us as if in a certain sense we evoke *sleep itself*, as Merleau-Ponty has described it:

> ... I lie down in bed, on my left side, with my knees drawn up; I close my eyes and breathe slowly, putting my plans out of my mind. But the power of my will or consciousness stops there. As the faithful, in the Dionysian mysteries, invoke the god by miming scenes from his life, I call up the visitation of sleep by imitating the breathing and posture of the sleeper....
>
> There is a moment when sleep "comes," settling on this imitation of itself

73. E. Jacobson, *Progressive Relaxation* (Chicago: University of Chicago Press, 1938[2]).
74. F.J.J. Buytendijk, *Ueber den Schmerz* (Bern: Huber, 1948), p. 58.
75. Kleitman, *op. cit.*, p. 113.

> which I have been offering to it, and I succeed in becoming what I was trying to be: an unseeing and almost unthinking man, riveted to a point in space and in the world henceforth only through the anonymous alertness of the senses....[76]

That falling asleep is not an activity performed by us, such as, for example, grasping or thinking, is already suggested by the personification of sleep [in the Netherlands sleep is represented as a man] : *he* comes, *he* overpowers us. We Dutchmen know a "sandman" (*Klaas Vaak*), that mysterious character whom all children have wanted to catch just as the moment he is sprinkling sand in their eyes and who precisely "attacks" when their attention relaxes. There are those nights, Gide says, in which sleep resembles a frightening animal which retreats any time one approaches it and which one nonetheless wants to tame.[77] In this mythology sleep becomes the other whom we encounter the moment we fall asleep and then lose again.

Whereas going to bed can indeed be considered an activity and the relaxation which accompanies it is started actively, falling asleep can no longer be called an activity. The one who rests maintains his own, "willed" attitude or posture and thus a relationship with the world from which resting precisely borrows its proper meaning: "it is only when oriented toward human work as the fulfillment of a *personal task*, that resting receives its genuine meaning, namely the consolidation of the proper starting attitude which as a *personal attitude* has the expressive character of something quiet and, merely through this, again means fitness for work."[78] While resting we may turn back in enjoyment to the comfortably relaxed body. The one who rests is "delightfully lazy" and in this he is in a dialogue with himself as relaxed body. He *keeps* quiet and is thus awake – until he begins to drowse and through this loses his body along with the world and himself. It is here that resting is essentially different from sleeping. Resting is a form of behavior, namely an active passivity in regard to the world which is taken up quietly by the one who rests. Falling asleep is de-activation. It is because of this character that we have no adequate term for this no-longer-behaving.

Whereas in resting there is a relation in regard to one's own body, or at least there can be such a relationship, in falling asleep this body is, as it were, dissolved. Arms and legs lose their place, orientation in regard to position is lost. It frequently happens that only the head is still experienced, particularly eyes and mouth, before these parts, too, are included in the de-activation. Sometimes it is also the hands which continue to "exist" although the arms have already gotten lost. For as long as there is attention, there are eyes; as long as there is a conversation, there is a mouth; and hands stay ready for grasping and handling until the one who is falling asleep gives up all grips. The organs "lose consciousness" in the order of their importance for our conversation with the world.

76. M. Merleau-Ponty, ***Phénoménologie de la perception*** (Paris: Gallimard, 1945), p. 191. (English: pp. 163-164.)
77. A. Gide, *Journal 1889-1939* (Paris: Édition de la Nouvelle revue française, 1948), p. 821.
78. Buytendijk, "Repos et sommeil." See for sleep and relaxation *De Vrouw* (Utrecht: Het Spectrum, 1951), pp. 232ff. by the same author.

When our body does not lose its incarnating function, but calls us back because of pain, an uncomfortable position, etc., we do not fall asleep. But again this is so only when the pain is understood and responded to as an appeal. Kant tells us how once having fallen prey to severe pains, he focussed his attention on "an indifferent object" through which the pains "became obtuse and thus sleepiness overpowered them."[79] By evoking a fading meaning he forgets his painful body via a detour and falls asleep.

Falling asleep means abandoning one's attitude. This holds not only for the body, but also for the whole person. "Attitude" does not refer merely to the taking of a position which we can see in the body, but also the person's taking of a stand which can be described also as an *inner attitude.*[80] We wish to consider once more in connection with the relaxation of attention this merging of the bodily and mental aspects. Attention itself is one of those phenomena psychologists have never been able to capture since it is by no means graspable as a thing. That is why Rubin has argued that attention does not exist at all: attention is a special way in which the field becomes structured,[81] or as we would say now the manner in which the world becomes centered as the focal point which draws our attention. However, Rubin is incorrect in conceiving of attention exclusively as a structuring of the psychophysical field on the basis of *Gestalt*-laws. Here, too, we do experience ourselves as origin and we observe in the other his attentive being-directed-toward.... That is why we should like to describe attention as an intentional relation between person and world which from the person's viewpoint can be determined as a fixing-oneself-on; and from the viewpoint of the world as a structuring of itself into figure and background. This attentive relation is pre-eminently a *reflexive* relationship, namely that through which we as human beings first acquire our human grasp of the world. *Attention is the original form of reflexion* and is described as such by Minkowski:

> Among the thoughts which are found more or less dispersed in the field of my consciousness, I focus my attention on that which, for one reason or another, must occupy me more particularly. Then the following images come to our mind: from a heap of thoughts all of which are more or less volatile, I grasp one, I fixate it, and try to maintain it. A parallel to the act of (manual) grasping establishes itself here. This comparison may appear gross, but that does not make it less natural, and as we shall see, it has a deep *raison d'être.* Is the word "main-tain" [*main* (hand)-*tenir* (to hold)] not proof of this?[82]

But whereas Minkowski in keeping with his "vitalism" sees attention primarily as a

79. I. Kant, *Von der Macht des Gemüts* (Berlin: Globus-Verlág), p. 66ff.

80. Cf. H. Lipps, "Die Haltung des Menschen," *Die menschliche Natur* (Frankfurt: Klostermann, 1941). Also: J. Zutt, "Die innere Haltung," *Monatschrift für Psychiatrie und Neurologie*, 73(1929).

81. F. Rubin, "Die Nichtexistenz der Aufmerksamkeit," *Bericht des IX. Kongresses für experimentelle Psychologie, München*, Jena, pp. 211-212. Cf. for this problematic: B. Petermann, *Das Gestaltproblem* (Leizpig: Barth, 1931).

82. E. Minkowski, *Vers une cosmologie* (Paris: Aubier, 1936), p. 89.

vital phenomenon, as that function which makes it possible to integrate such "static" and "dead" elements as the perception of unmoveable things and our thoughts about it into life,[83] we see in attentiveness taken as the basic form of reflexivity the original human phenomenon through which vitality precisely becomes elevated to a human vitality. But we do not wish here to go into the philosophy of reflexion which should sometime try to re-evaluate this "adversary" ("*Widersacher*").

Attention is that which interrupts the stream of experiences and splits it open. We agree with Minkowski on this point. When he determines the essence of attention as "dwelling-on" (*s'arréter-à*), then the bodily and mental signification of the dissociating relation with the world is contained therein. To listen, to see, and to think attentively are forms of "dwelling-on" no less than shortening one's step. No wonder that Ribot wrote: "The actions of coming to a stop appear to play an important but still poorly known part in the mechanism of attention."[84] Our consideration which was oriented sometimes to the body sometimes to the mind, must become a consideration of the *person*. To say something about the body is tantamount to saying something about the mind and vice versa.

Can it then be amazing when Jacobson notes that sleep comes to the extent to which our relaxation increases and particularly when our *eye muscles* relax?[85] For the eye is pre-eminently the instrument for all *fixation*, for fixing-oneself-to and confining-oneself-to..., just as the *eye of the mind* is the mythological instrument of reflexion. Jacobson remarks at the same time that *attention* slackens by increasing relaxation[86] and that paying attention to one's own movements is detrimental to relaxation.[87]

However it would be incorrect to try to understand falling asleep exclusively from the point of view of our slackening attention. We have pointed repeatedly to the difference between the reflexive level and that of our pre-reflexive experience. As for the latter, there, too, a certain change must occur before falling asleep becomes possible. Whereas in the first phase of falling asleep withdrawal of the grasping intention is the dominating characteristic, yet in a later phase the unreflexive conversation with the world must be silenced, also. As we have seen, our experience receives a certain autonomy when we eliminate the reflexive taking of a position. We say a certain autonomy, in that the harkening-back-to never disappears completely, not even during a "pure experience." That is why there is the possibility of returning to the experience of falling asleep and analyzing it – through which it naturally loses its experiential character immediately. Even in the dream where there barely seems to be a question of reflexivity, we are able to return *freely* to the state of waking when we do not like our dream.

Sartre has made some remarks on this "wanting to awaken." The only means of

83. Minkowski, *op. cit.*, p. 94.
84. Th. Ribot, *Les maladies de la volonté* (Paris: Alcan, 1916[29]), p. 106.
85. Jacobson, *op. cit.*, p. 297. Cf. also: Sartre, *L'imaginare,* p. 60.
86. Jacobson, *op. cit.*, p. 111.
87. Jacobson, *op. cit.*, p. 46.

getting out of his dream at the sleeper's disposal, is the reflexive realization: *I am dreaming*. But in order to achieve this return to one's own dream-experience, reflection is already pre-understood. Sartre says rightly: "But this reflexive consciousness is almost impossible to produce because the types of motivations that ordinarily call it forth are precisely of the sort which the "enchanted" consciousness of the sleeper no longer permits itself to conceive. In this connection nothing is more strange than the desparate efforts made by the sleeper in certain nightmares *to remind himself* that a reflexive consciousness is possible." And he notes that a shocking emotion is necessary through which this harkening-back-to is motivated.[88] Although it is true that only penetrating analyses can provide us with clarity, there cannot be any doubt that the possibility of reflecting is given in principle in sleep also, and certainly in dreams.

And this means that we never sleep through and through. But then perhaps "complete sleep" is a mere theoretical construction, a limit-idea. When in my sleep I-in-my-origin have slid back into an almost plantlike mode of existence, then still a last, extremely vague kernel remains ready to unfold itself again into a living center of interests at the slightest signal. In my sleep I have not disappeared; *I* am sleeping, and in so doing remain ready to show my originality. Sleep is a state in which I, *anonymously*, just merely live, but am continuously ready to wake up as I-myself. Thus we reverse Klages's thesis because only then does it become adequate: *in the passivity of sleeping there is continuously hidden a readiness for the activity of waking-up*. That is why sleeping strictly speaking is never *absolute* sleeping but a continually renewed *falling-asleep*, that is an abandoning of activity – until inexorable light calls me back and I-myself along with my world wake up from anonymity in order to continue our conversation.

Why do we speak here of a renewed falling-asleep? Because the sleep in which we are submerged is only seldom a stable *state*. First there is the continuous readiness to wake up as soon as an inexorable appeal is realized. But further it is true, also, that the "inexorability" of the appeal which wakes me up already rests on a "decision" on my part. While sleeping I "deal" with a number of disturbances as being unimportant; they reach me vaguely because I am still "somewhere" awake in an anonymous way; but I *disinterest* myself from them and that means that I renewedly fall asleep.

VII. THE SITUATION OF FALLING-ASLEEP AS PERSONAL SITUATION

I wake up *with my world*, just as I slept, *personally*. I-in-my-origin do not wake up in pure originality, but in a continuation of my history. It is in this respect that my *re-birth* in waking-up is distinguished essentially from my original *birth*. My past is not erased by sleep; the future which lies before me is not an absolutely new beginning. Falling-asleep and sleeping are integrated in my history which manifests itself in the peculiarities of these phenomena, also.

88. Sartre, *L'imaginare, p. 224.*

Sleep is undoubtedly a "biological phenomenon"; but not a "foreign object" in our personal existence which stricity speaking, has nothing to do with it. The structure of the world proper to falling-asleep necessarily yields data concerning the history of the person in question. In our analysis of falling-asleep we have repeatedly encountered this being-integrated in a meaning-history. Now at the end of our investigation we wish to make this once again into an explicit theme for reflection by giving a concrete example. The general structure of falling asleep interests us here, then, merely insofar as it manifests and concretizes itself in a personal existence. To show this we have chosen the sleep-conflict as Proust described it. The author tells us that for a long time he was accustomed to going to bed early. The candle was barely snuffed when my eyes would close, so quickly that I had no time to realize: now I am going to sleep. Half an hour later the idea that it was getting time to go to sleep woke me up. In other words, he had not yet slept at all; he drowsed and after the withdrawal of the reflexive acts he moved into a stilled world of experience in which he no longer distinguished *himself* from the world because the reflection on his own experience failed to come. That is why he says: while sleeping – that is half asleep – I had not stopped to concern myself with what I had just read; but, these "thoughts" had taken a strange turn: "I myself seemed actually to have become the subject of my book: a church, a quartet, the rivalry between François I and Charles V. This impression would persist for some moment after I was awake; it did not disturb my mind, but it lay like scales upon my eyes and prevented them from registering the fact that the candle was no longer burning."[89] This identification of the experiencing-I with the subject matter of the dream takes place through the "autonomization" of experience which is recaptured in reflection not at that very moment but only later in memory; I-myself let myself glide and am not *authentically interested* in my experience; there is no reflection and this manifests itself through the fact that the *eyes* do not realize that the candle no longer burns.

The interest which was stilled now returns and *alienates* him from the world of his experience: Proust describes how his dream-world becomes unintelligible for him; the event *separates* itself from him, he is free to choose whether he will form part of it or not. This is to say, reflection returns, but it is still "free" and "free-floating." I have not yet "decided" to take up the appeal of the world evoked in my mood; whether I shall enter it unscrupulously or whether I shall wake up completely; whether I shall unfold myself into a knowing I, or whether I shall keep floating dreamily. If I wish to *dwell* in this world which is unclear and darkening, create and encounter my own figures in this amorphous world, and yet would really go to sleep, also, this is the point where a conflict will arive; there are so many possibilities. I can slip back into sleep, let myself be awakened by the real world, let myself float in the stream of my experiences; or finally – and this is what Proust chooses – dreamily turn back to myself and can then try to encounter "the" ego in its

89. Proust, *op. cit.*, p. 9, (English: p. 3). Cf. E. Lévinas, *De l'existence à l'existant* (Paris: Fontaine, 1947), p. 111: "In insomnia there is not *my* vigilance in regard to the night, but it is the night itself which watches. That stays awake. In this anonymous watchfulness in which I am completely exposed to being, all thoughts which fill my insomnia, are suspended *to nothing*."

originality. That is why Proust rejects the phosphenes which force themselves upon him and alienate him from the world of experience in which he thinks he can find himself:

> And, indeed, I found plenty of charm in these bright projections, which seemed to have come straight out of a Merovingian past, to shed around me the reflections of such ancient history. But I cannot express the discomfort I felt at such an intrusion of mystery and beauty into a room which I had succeeded in filling with my own self *until I thought no more of the room than of my self.*

This room is he *himself.* He is now really awake, his body has turned over once more for the last time and the angel of certainty had kept everything around him steady the way it was while he was asleep: himself under the blanket, the room, the chest of drawers, the desk, and all the rest. But of what use is this reality for him? Of what avail is this *knowing* that he is not *there* where he thought himself to be at the moment between dream and reality? But his memory responds to the appeal of what is past and forces itself upon him. Usually I did not wish to immediately fall asleep again at all, Proust says, but spent the greater part of the night recalling my past. He abandons this knowing-waking and turns to the country of his origin, to his origin. But this turning is already reflexive: "I started to think, experience, *things so sad.*"[90] But thinking and experiencing, too, alienate me from my source, from the genuine I which Proust wished to *see*. That is why Briand says that Proust asks his room not only to isolate him from the outer world but also from the inner world which is projected in thought and experience.[91] Proust does not want to sleep, but to find himself. He experiences everywhere the alienation of his origin and thus asks for the land of his youth. Thus, this is our first point: Proust cannot surrender to sleep because he is not secure. But he surrenders to dreaming, to the "waking sleep" because he loses himself while being awake; and this is the second point. Waking *and* sleeping are spoiled for him. Here we find that remarkable ambivalence we also encounter in Rilke, Baudelaire, and Amiel. Insomnia, and in general sleep-conflicts, also imply waking-conflicts. It is impossible for one to live a harmonious life during the day and be unable to go on at night. In these cases we see the veneration of sleep connected with insomnia; a horror for waking combined with a continuous being-awake. Proust does not want to fall asleep, but he will not

90. Proust, *op. cit.*, pp. 13ff.

91. Cf. Briand, "Maladie et sommeil chez Proust," *Les Temps Modernes*, 51(1950), p. 1179. This article was reprinted in *Le secret de Marcel Proust* (Paris: Gallimard, 1950), pp. 18-52. Note what follows in the passage quoted: "His world, the only one in which he feels at home, in which he finds himself again and recognizes himself, that is the pure interiority, that is the not-formulated, immediate, intuitive and, as it were, unconscious apprehension of the ego by the ego. It is the sudden shock of life in its own source caused by this being which is born in it taken at the moment when this is being born in it, independently of every fact of consciousness which could not do anything but, by exteriorizing it, dissolve the one as well as the other, being and life." See for the "failure" which must assert itself with this: G. Gusdorf, *La découverte de soi* (Paris: Presses Universitaires de France, 1948).

not be awake, either; he will both, and yet he will neither one of the two. In his sleep he is threatened, but being awake he is uncertain. Neither while awake, nor asleep, does he encounter the world as a milieu in which he is at home and safe. This means that the self-reliance of his existence in a mature, threatening and yet reliable world is not accepted by him. The turning back toward a lost youth and lost ego are inseparably connected with this. Proust does not accept the fact that youth and ego in their originality must be "lost" of necessity.

In the sleep-conflict as he describes it for us we see how it is motivated by *the absence of his mother*: he is *unable* to sleep as long as his mother has not come to see him in his bed and he longs for the moment in which she will come. It is this longing that keeps him from falling asleep. But once she comes upstairs it means that later on he will not be able to fall asleep because she will have left him again; that is why he hopes that it will still take a long time before she comes.[92] However, when she does not come he sends her a short letter saying that she must come. But the mother has guests and is long in coming. This is why Marcel closes his eyes and attempts no longer to hear the voice of his parents. But he does not succeed in falling asleep:

> ... I realized that, by writing that line to Mamma ... I had cut myself off from the possibility of going to sleep until I actually had seen her, and my heart began to beat more and more painfully as I increased my agitation by ordering myself to keep calm and to acquiesce in my ill fortune. Then, suddenly, my anxiety subsided, a feeling of intense happiness coursed through me, as when a strong medicine begins to take effect and one's pain vanishes: I had found a resolution to abandon all attempts to go to sleep without seeing Mamma....[93]

Is the issue here about his mother? Or does she merely represent the feeling of security Marcel is unable to find either awake or asleep? At any rate it is the canvas of the personal history which determines the situation of falling asleep, or as is the case here, the structure of the sleep-conflict. Inversely, the inner life history of the person betrays itself here in this conflict as the center of significations.[94] The neurotic sleep-conflict will seldom have found its equal so adequately as in the case we have here with Proust. The insomniac *cannot leave the world*, is unable to sur-

92. Proust, *op. cit.*, p. 18.
93. Proust, *op. cit.*, p. 33.
94. Cf. Proust, *op. cit.*, p. 51 the description of the insomnia cult of Aunt Leonie who has lost her husband and therefore no longer "can" nor is allowed to sleep because of sorrow. "Unfortunately, having formed the habit of thinking aloud, she did not always take care to see that there was no one in the adjoining room, and I would often hear her saying to herself: 'I must not forget that I never slept a wink' – for 'never sleeping a wink' was her great claim to distinction, and one admitted and respected in our household vocabulary; in the morning Françoise would not 'call' her, but would simply 'come to' her: during the day when my Aunt wished to take a nap, we used to say just that she wished to 'be quiet' or to 'rest'; and when in conversation she so far forgot herself as to say 'what made me wake up', or 'I dreamed that', she would flush and at once correct herself." (English: pp. 70-71.)

render, does not know of relaxation and resignation, but he must be *with* it and thus be awake. He knows no rest. He turns toward the past because he does not feel at home in the present; and he does not feel at home in the present because he turns back toward the past. He desires to sleep in order to escape from being awake; but this desire precisely keeps him awake. Nobody is more caught up in reflection than he who tries to escape it. To wake means not only to be awake, but also being wakeful, being careful, paying attention, letting nothing escape one, being tense – and thus being active, directed-toward, being busy with, being in conversation with. In a word to wake is "*sich auseinandersetzen*".

I am asleep and *I* am wake; and it is in the fluctuation of these two that my history unfolds itself as *mine*. We may lose sight of this personal meaning and determination of the situation proper to falling-asleep in analyzing *falling asleep as such* as a relation between person and world in general. For a psychologist this analysis is a preliminary phase which he must go through before he comes to his real task which must be defined as the return to the concrete person. The psychological character of the problematic which oriented this investigation implied that we had to limit ourselves to the person. From the perspective of a pure phenomenological analysis the preceding reflections remain caught up in a psychological naivité which would have to be overcome before the genuine act-analysis in regard to I-in-my-origin could start. Only then would a foundation of our analysis in the general theory of intentionality be possible, also. That is why this contribution to phenomenological psychology is a provisional investigation from the viewpoint of pure phenomenology.

When we comprehend falling-asleep as an abandoning-relaxation of the person, that is of body and mind, as a dying out of intentional acts and a becoming still of their correlates, then all the phenomena we encountered in the course of our investigation must be understood from that view. We *cannot* fall asleep; sleep lies outside the possibilities of our will; being able to fall asleep does not flow from a "capacity", nor from a mysterious "*vis dormitiva*", but from the readiness to sink back to being just merely a living body which is nevertheless still somewhere awake in a mysterious way insofar as our interests maintain a grip on our sleeping interest. We cannot observe and fix our experience of falling asleep without being with it and thus without remaining awake.

In falling asleep the issue is not that we *turn away from* something; all aversion is at the same time a *conversion* to something else. He who falls asleep *ceases* to turn to something. Each activity is a grasp on a world which invites him to activity; both, activity and world, must fall asleep together; the world by becoming still and by manifesting itself no longer; the person by surrendering to disinterest.[95] I shall "let come and go whatever may happen"[96] and fall asleep.

95. Bergson (*op. cit.*, p. 95) says: "... suppose that at a certain moment *I wish to be disinterested* in my present situation, the pressing action, and finally in what concentrates all the activities of memory on one single point. In other words, suppose that I fall asleep." However, cf. Claparède, "La question du sommeil," p. 434, concerning the relationship between sleep and

Falling asleep is the becoming silent of a conversation. There is no longer anything which speaks to me, or if so, it is merely in a way that I no longer need to respond. I, on my part, keep silent and go to bed and thus unintentionally and unnoticed leave a world which can no longer hold me and call me to myself – in order to wake up tomorrow as a reborn man.

How must we do this? By doing nothing. All methods of falling-asleep amount to this: find access to sleep which is not intentionally willed by means of a willed fading of all appeal, by means of a de-activation of our activities.

However, the essence of falling asleep is not contained in the *terms* of this analysis. This analysis has reached its goal only if it is continuously confronted with the phenomena and is confirmed in a direct intuition.

May I therefore be permitted to evoke a final image in which once again is implied what we have wished to make explicit: the water waits until the wind dies down; then the ripples will slide away all by themselves, and the water will become smooth and sleep in its depth. If we accept this image as an *image* of falling asleep and no more, then, perhaps we may admit, also, that no one can evoke such an image for us better than Edgar Allan Poe in his description of the water that falls asleep:

> And the lulling melody that had been softer than the wind-harp of Aeolus, ... it died little by little away, in murmurs growing lower and lower, until the stream returned, at length, utterly, into the solemnity of its original silence.[97]

APPENDIX

EDMUND HUSSERL: *THE UNCONSCIOUS I – SLEEP – IMPOTENCE*[98]

The I does nothing of all it could do and its being-able-to is no longer "alive"; it does not hold to anything to which it usually holds fast, the "retention" is becoming absorbed without a holding-on-to, without interest on the part of the I; the

---

the "law of interest"; also in "Le sommeil et la veille," p. 448. Allers, too, relates (subjective) tiredness and disinterest on the one hand, to "the capacity for sleep" (*Schlaffähigkeit*) which is motivated by them: "When someone during some work or other suddenly loses his interest in it, for instance realizes that he cannot beat his competitor, then sometimes fatigue sets in suddenly.... In fact there are attitudes which favor the occurrence of such experiences, as for instance lack of interest, the conviction that a work is difficult and impractical, inner rejection of this work, and so on." Cf. R. Allers, "Ueber neurotische Schlafstörungen," *Deutsche medische Wochenschrift*, 54(1928), p. 817.

96. Jean Paul, *op. cit.*, p. 84.

97. E.A. Poe, *Eleonora*, in *The Complete Poems and Stories of E.A. Poe*, vol. I, (New York: A Knopf, 1946), p. 375.

98. After completing this essay we found in the Husserl-Archives at Louvain the fragment *Das bewusstlose Ich – Schlaf – Ohnmacht* which we here publish in its entirety because of its agreement with our own analyses. The author thanks Professor Dr. H.L. Van Breda O.F.M. for his kind permission for publication.

horizons which are already formed become non-living horizons, they are not affective horizons, the interest is lacking, that which a moment ago was still awakened by an interest from an association that was still functioning, no longer has any influence on the I, it does not orient itself toward it; now it no longer has any orientation in regard to anything, it is even without any interest in anything, it is without any specific "consciousness-of" (*Intention*), it sleeps, it is "unconscious." The unconscious I is in Nirvana; its willing and doing is a dying of all interest, it is moved by nothing, that is to say, it is moved as something which is not moved in its interest; as something which is without any interest, it does not move, it does not do anything, it does not experience, it does not see anything, hear anything, accomplish anything, etc. However it can wake up – thus we have here two modi:

1. There is first the consciously active I, I in the genuine intentionality; the I which is oriented-toward something via acts, but which has a horizon of existing objects towards which it is oriented implicitly, a world-horizon, world taken as the world of the ego's articulated interests.
2. The other modus is the "unconscious", or sleeping I, which in this mode has no interest. However the waking and interested I is the same as that which is now without interest, and in this sameness it is also interested in the world which formerly had become without interest for it as that world which, in its earlier interest, it itself had constituted. The constitution and what is achieved by it are not lost, but it has fallen into the modus of submersion as interestlessness, and from the side of the I into a "pure" sleep, which would be equal to the impotence of unconsciousness, absolute "act-less-ness". The awaking of the I is the awaking of the center of interest, of genuine intentionalities. That is why we should distinguish, it seems, between an affection that awakens the I and affections which the I, already awake, experiences – just as it wakes up, so also all its interests wake up immediately or mediately, the association movements as propagations of materialized interests are oriented towards something which is already in whatever modus it may be as something which is suitable for its living interest. The entire world is awakened (it is not so that just what is of prime interest becomes isolated), which is for *me* (just as it has value for me), and now from my living interest beams go out into that world waking up special interests, and this awakening is a making-stand-out and a motive for associative reproduction.

Certainly we should also reflect then on the difference between increase and decrease of interest – thus on modi of the form of being-awake, and how decrease

---

The fragment consists in an appendix of two typed pages (pp. 48-49) which refer to pp. 17ff. of the (transcribed and typed) ms. A. VI 14 which is entitled: *Die phänomenologische Problematik von Geburt, Tod, Unbewusstsein zurückgeleitet zur allgemeinen Theorie der Intentionalität. – Weltbewusstsein und thematisches Bewusstsein.* The *ms.* dates from the years 1930-1932.

On pp. 10ff. of the main text Husserl discusses the problems of sleep and falling asleep in a broader context. That is why these pages, however interesting they may be, are not so suited for separate publication as the presently reproduced appendix which forms a complete whole.

of interest as decrease of being-awake can lead to falling asleep. Thus the issue here is that of our total interest in which manifold special interests play a part and determine the course of our active life, as a course of a real, intentional temporalization. Time is already temporalized, it is already time-world, but the really original and ever continuing temporalization in the development of the real constitution of the world, presupposes a life which is awake and materializes itself in the form of a life of interest.

6

The Phenomenological Approach to the Problem of Feelings and Emotions*

F.J.J. BUYTENDIJK

During the many years of my investigations of animal behavior I have become more and more convinced that we ought to understand the observable vital phenomena such as actions and expressions. Behavior can never be reduced to physiological processes and explained as a result of the integration of reflexes. The reflex is a reduced action, stabilized by the constant signification, for example the dangerousness, of the situation in which it occurs. The characteristics and the entire signification of behavior become intelligible in the light of their relation to the essence of the animal being in general and the definite mode of existence of the species.[1] The proper content or signification of the concept of behavior presupposes that being an animal is absolutely different from being a crystal, a drop of water, or a plant.[2] These ideas and views resulting from concrete experiments and observations have made me appreciate the value of the phenomenological approach and apply the phenomenological method to the problems of psychology.[3] Animals and men are observable subjects[4] because we understand behavior as a system of intentional acts.

Husserl's phenomenological method[5] is founded on the irreducible fact that consciousness must necessarily be consciousness of something toward which it is intentionally directed and which therefore has a meaning. Consequently consciousness is

* From Martin L. Reymert, ed., *Feelings and Emotions*. The Mooseheart Symposium. (New York: McGraw-Hill, Inc., 1950), pp. 127-141. Reprinted by permission of the publishers.

1. F.J.J. Buytendijk, *Wege zum Verständnis der Tiere* (Zürich: Niehans, 1938).
2. This is explained in an excellent way by Merleau-Ponty in *The Structure of Behavior* trans. Alden L. Fisher (Boston: Beacon Press, 1963).
3. F.J.J. Buytendijk, *Das Spiel von Mensch und Tier* (Berlin: Wolff, 1933); *On Pain*, trans. Eda O'Shiel (Westport Conn.: Greenwood Press, 1973); *Algemene theorie der menselijke houding en beweging* (General Theory of Human Posture and Movement) (Antwerpen: Standaard Boekhandel, 1948).
4. Victor von Weizsäcker, *Angeschaute Subjektivität* (Leipzig: Der Gestaltkreis, 1943).
5. A very good survey of Husserl's ideas is to be found in Marvin Farber, *The Foundations of Phenomenology* (Cambridge Mass.: Harvard University Press, 1943); Marvin Farber, ed., *Philosophical Essays in Memory of Edmund Husserl* (Cambridge Mass.: Harvard University Press, 1940). A short but very clear explanation of the concept of intentionality has been published by J.P. Sartre, "Une idée fondamentale de la phénoménologie de Husserl: L'intentionnalité," in *Situations*, 6 vols., (Paris: Gallimard, 1947-1965), vol. I.

not considered to be an "interiority," an imaginary space with certain contents and processes, interacting with the physical being and with a sensorially appearing outer world. This conception, implied in various images, has entered psychology as a means of interpreting consciousness (or the self or the ego) as "something" existing (*res cogitans*) that enters into relationship with an existing world (*res extensa*).

Psychology as a positive science has become dogmatic and lacks the radicalism of doubt. The unprejudiced return to the "things themselves," *i.e.*, inspection of knowledge "itself" as given to us *directly*, with nothing mediating or interfering, reveals the intentionality or act-character of all behavior. Feeling, *e.g.*, being pleased, is an act, in which there is reference – not a causal relation – to an object that is intentionally *present*. It may occur without the object existing, for this object is merely *meant* in the act of feeling.

Perhaps, someone might remark, psychology as a pure science has nothing to do with a theoretical conception concerning consciousness, neither the Cartesian nor the phenomenological idea. Science should be limited to establishing *facts*. I can agree with this conception of science, but then, I think, we are obliged to ask what a fact is. Is feeling a fact? Or is it an abstraction? And is it a fact that there exist men who have feeling in certain situations? If we take this question seriously, we should have to know what we *mean* by "exist," "situation," "feeling." This is exactly what it is the aim of phenomenology to reveal.

When we should like, for example, to institute a statistical inquiry into "happy and unhappy marriages," we have to know *what* a marriage is. We can arrange to take it for granted that a marriage can be defined by certain characteristics or properties, just as we identify a substance as oxygen or gold. I will not insist on the dubious value of such an objective determination of a human relationship such as marriage. But what are happiness and unhappiness? Should we calculate in our statistics the subjective opinions of everyone who is asked whether he is happy or unhappy or should we try to know what may be the essential characteristics of the sense of happiness as a pure phenomenon, given to us directly as something that shows itself in itself, with all its immanent characteristics and also its reference to the situation of being married? This analysis of the intentional experience is the phenomenological approach to feelings and emotions.

In phenomenology we should like to know what the significations of the acts of hating, loving, desiring, rage, joy, etc., are; and we are convinced that these significations or meanings are the real essence of the feelings. Every feeling is a feeling of something, and the human attitude in which a feeling is experienced in a positional, not reflective, consciousness results in fuller understanding.

Our feelings are not senseless states of consciousness or psychic facts, but modes to detect the signification of situations, to *know* what is savory, disgusting, alarming, distressing, lovely, etc. Human reality is equivalent to being conscious and this can be defined only as open to the world, as cast upon the world. Being conscious is *s'éclater dans le monde*, to use Sartre's image of this kind of being. A pure description of purified knowledge of the affected and feeling self, of the situation and the concrete mode of existence in this situation, ignores all actual facts, for example, psychogenetic laws and physiological conditions.

In phenomenological psychology man is considered as existing. All mental experiences – perception, remembrance, thinking, dreaming, and also feeling – are relations of an existing human being and his world. They presuppose both the human being and his world, and we ought to know what is meant by these concepts. To discover the definite significance, *i.e.*, the essence of a feeling, we have to start from a conception of hjman existence. Consciousness exists to the exact extent that it is consciousness to exist. It would be a misunderstanding to think that the a priori concept of existence can be interpreted as a dogmatic a priori, that is, a statement about human existence of irrefutable validity.

Surely all phenomenological analysis of feelings and emotions, such as Scheler's investigations of the modes of sympathy,[6] the research on normal and abnormal feelings by such psychiatrists as Rümke,[7] Binswanger,[8] Minkowski,[9], Kunz,[10] and others, and the analysis of laughing and crying by Plessner,[11] presuppose a certain implicit idea about human reality and man's existence in the world, but this conception becomes explicit in the progress of the phenomenological analysis itself. If the psychologists in their modesty are animated by the laudable hope of realizing later, on the solid ground of their research work, an anthropological synthesis, they are in full contradiction with themselves (Sartre[12]). The traditional theories of feeling and emotion, even all descriptions of these phenomena, are based on the a priori conception of man as a "mundane ego," that is, a constituted object among other objects in the world. We do not agree with this conception.

The phenomenological approach to feelings and emotions starts from the undeniable fact that consciousness is always a being conscious of something else and that we are conscious of our existing, that means our being physically subjected to a given situation. This "being subjected" is not a causal relation, but it means that the situation must be responded to. The reply is the attitude, the feeling, the intentional act and not a reaction in a physiological sense. The fact that we find ourselves as subjects (in our "this-ness") only by standing apart and viewing ourselves in the situations in which we are involved is the profoundest basis of the phenomenon of feeling. Feeling and emotion are the affirmations of our attitudes toward situations, and the pure phenomenon of feeling reveals the human being as always projecting it and always projecting the world. The pure description of a feeling is the description of an existing human being in his well-defined attitude toward a

6. Max Scheler, *Wesen und Formen der Sympathie* (Bern: Francke Verlag, 1973).

7. H.C. Rümke, *Zur Phänomenologie und Klinik des Glücksgefühls, Monographien Gesammtgebiet Neurologie und Psychiatrie*, 39(1924).

8. Ludwig Binswanger, "Über die manische Lebensform," *Schweizer Medische Wissenschaft*, 75(1945).

9. Eugène Minkowski, *Vers une cosmologie* (Paris: Aubier, 1936).

10. Hans Kunz, *Die Aggressivität und die Zärtlichkeit* (Bern: Francke, 1946).

11. Helmuth Plessner, *Laughing and Crying: A Study of the Limits of Human Behavior*, trans. James Spencer Churchill (Evanston: Northwestern University Press, 1970).

12. Jean-Paul Sartre, *Sketch for a Theory of the Emotions*, trans. Philip Mairet (London: Methuen and Company, 1962). This book is also available in the form of a translation by Bernard Frechtman: *The Emotions: Outline of a Theory* (New York: 1948). The passage referred to is to be found in this edition on p. 5.

situation. The projection is the signification of the act of feeling, a spontaneous movement (*s'éclater vers* – ejecting himself toward), which implies a totality of relations and their development. Feeling is a mode of replying to a situation and transforming it as a projected new world, in which unknown qualities are categorically experienced. There are as many feelings as there are situations, and the situation is created by the mode in which I have accepted it, *i.e.*, by my chosen projection. Of course, this projection is not the result of reasoning and my choice is not arbitrary. I choose my emotional attitude in the same unreflective, nonconsidering (thematic, not thetic) mode of consciousness as I choose my words in speaking or writing.

The attitude toward the situation, confirmed by the feeling, is a true reply that is elicited. It is always my own unreflected act. There are, for example, many kinds of being irritated, angry, or furious, in which we choose and project ourselves in the mode of an efficacious revolt, just as I choose the more tasty of two apples. I have no motive in mind for my choice, but still there is in the situation and the backward reference to my existential history an antecedent that works *only* by its signification and hence is no cause.

When the normal impassive contemplative attitude becomes too difficult, so that we cannot meet the exigencies of the situation, we may choose to become angry and to change the situation and our attitude to it by this feeling.[13] Every feeling has a signification that makes us find the thing that is signified. Both the dynamic quality of the immediately experienced feeling and the essential characteristics of the expression refer to the signification and the thing signified. Even joy is not an automatic reaction to an impression. Joy may be cheerfulness, high spirits, radiant expansion, dilatation, illumination, elevation.

All these dynamic elements may be the invariant content of a certain kind of joy. They denote the act, by which I affirm my choice of the positive value concealed in the situation and now revealed by my feeling joyous. The phenomenology of feeling has taken only the first step on the way of analyzing the various kinds of joyous feelings. We know – and it is important for psychopathology – that there exists a silent, quiet joy, large, placid, voluminous, embracing as a climate and signifying the mode of possessing and concealing a value which structures our entire existence.

There is also a leaping joy, a being animated, when we shout, jubilate, dance. This joy is allied to and impregnated with impatience (Sartre). It signifies the anticipation, a reaching, longing for some value that is approaching, that is expected.

I will try to give a provisional short sketch of a possible phenomenological genetic analysis of feeling and emotion, starting from the conception of Ruckmick[14] that the affective life begins with consciousness itself. In every concrete existence this "being conscious" is consciousness of existing physically, *i.e.*, being limited. The pure phenomenon of existing is the immediate act of experience of its own

13. *Ibid.*, p. 22.

14. C.A. Ruckmick, *Psychology of Feeling and Emotion* (New York: McGraw-Hill, 1936), p. 214.

limit and this experience includes necessarily the experience of a possible transcendence of this limit. Conscious existing contains (in its own limitation) the possibility of feeling in its original signification of close relationship to the act of touching.

The sense of touch is the "father" of all senses (Lavelle,[15] Nogué[16]). Touching is the discovery of our real existence in its own limitation. It reveals at the same time the physical self as touched, moved, and something that is touched by our self-movement. Touching is the most original mode of the experience of participation and feeling. In this encounter, for example, feeling the smoothness, slipperiness, stickiness, angularity, roundness, the shape and form of an object, we are intentionally involved in an anticipated developing situation. The act, the answer to the encounter, may be in the direction of the discovery of the qualities felt or of the feeling of something or in the direction of being emotionally stirred up and disturbed. It depends on the situation and our approach to it, whether feeling or emotion will prevail. Every new encounter is at the same time the abolishing of a former mode of being-in-the-world and this former existence is stirred, moved, and emotionalized. Awareness of emotion and also of excitation, I believe, is a nonintentional element of every feeling, just as it may be an element in the act of touching. It depends on the situation and the projection of our mode of relation to it, if feeling, emotion, or excitement prevails. The original signification of the act of feeling and the disturbing function of emotion and irritation, which are included in touching, the most original act of being-for-self in the individual's limited physical existence, inhibit themselves in the manifold situations in which a human being chooses to be engaged. These situations are the affective ones which elicit rage, fright, terror, astonishment, sorrow, joy, shame, jealousy, hate, sympathy, compassion, etc., and unlimited variations of sentiments, passion, and moods. In all these cases the phenomenological analysis will have to detect the full content of the significant formation and the essential history of significance. Many such investigations have appeared in the literature. Typical examples are given by Sartre, who showed the magical transformation of the world by feeling and emotion. He also speaks of the projection of the affective significations on the environment and mentions that consciousness "lives in the new world which it has just constituted. Consciousness is the victim of the trap it has set itself. Consciousness is moved by its own emotion."

Emotion is not an intentional act, but allied to sensation, irritation, and excitation. We may come to understand this fact in the following way: we have only consciousness of ourselves (regaining ourselves in the vision of things). When I admire something or hate someone, I, in *one* intentional act of feeling, project both myself as admiring or hating and the qualitative structure of the object or person. The created situation is not a static pattern, but an animated, vivid, lively unity that appeals to me by the claims of its qualities, provoked by my feeling. As has been

15. Louis Lavelle, *Dialectique du monde sensible* (1921) (Paris: Presses Universitaires de France, 1954).
16. J. Nogué, *Esquisse d'un système des qualités sensibles* (Paris: Presses Universitaires de France, 1943); *La signification du sensible* (Paris: Aubier, 1936).

demonstrated by V. von Weizsäcker for the act of touching, there is a functional circle of movement and being moved, in which the *Gestalt* is developed. Exactly as in touching, there is in the act of feeling a *closed* reciprocal relation between the subject and its intentional object. Consciousness is shut up with something and isolated from the rest of the world. The *rebound* of my projection of feeling against the created intentional experience has the character of being moved, *i.e.*, of emotion (possibly in the modality of excitation, excitement, or irritation). The result is penetration of feeling with emotion, of emotion with feeling, a stabilization, a continuity of the affected relation, with a certain analogy to the standing vibrations in a closed resonator or the standing waves on a limited surface of water. Thus we conclude that there is no emotion without the act of feeling, but the emotion itself is not an intentional act. It is the specific quality of our own existence, revealed in the regaining of ourselves in the act of feeling.

This may be the reason that I prefer to say: *I am* admiring this picture, *I am* hating that person, and I am moved *by* the beauty of the picture, *by* the untrustworthiness of that person. We must be aware that in every language the verbal expressions are ambiguous. The reason is that the dialectic and reciprocal relation of the involvement of affect makes it possible to describe the experience in a different way. It may happen that the repercussion of the (created) object to my act of feeling is so intense that I am conscious of my existence as a purely-passive-entity-being-moved and that this emotion totally hides the original intentional act that was the creator and also is the maintainer of the emotion. Therefore Sartre could write "Pain is quite free from intentionality."[17] This is true, but there is no pain without a preceding, original attitude toward our existence as a physical-being-in-the-world. In our well-balanced life this attitude of ours is so "normal" that we forget its intentional character, but the simple fact of being painless in fighting or in hysteria reminds us that the condition of undergoing the sensation of pain is to be *with* ourselves, means to be conscious of our existence. Pain is not a feeling (act) but an emotional sensation only possible in the intentional experience of existence.

Continuing our sketch of a genetic analysis of feeling and emotion, we ought to realize that the encounter of our existence with something that limits it takes place in different intentional experiences. There are, so far as I can see, two extreme possibilities. The intentional object to which we are directed in the act of recognition of our own limitation is presented either as a resistance or as no resistance. The experience of resistance signifies our obstructed existence. The experience of no resistance in the encounter signifies our expanding existence. It seems to me that in grasping the sense of these phenomena we can distinguish four modes of feeling pleasant and four modes of feeling unpleasant, corresponding to the experience of different modes of existence. The four modes of existence that are presented in the feeling of being pleased are (1) being-with-something, that is, being one with something or someone (*e.g.*, in love), (2) flowing-on, flowing-out existence, (3) expansion of the accentuated self, and (4) taking to one's self, assimilating, including as

17. Jean-Paul Sartre, *L'être et le néant* (Paris Gallimard, 1948), p. 398; English translation by Hazel Barnes, *Being and Nothingness* (New York: Philosophical Library, 1956), p. 328.

one's own. These four dynamic contents are in outline four kinds of feeling pleasant which are involved in the most different concrete feelings actually experienced in life.

The encounter with a resistance has also a fourfold typical dynamic structure, corresponding to four modes of unpleasantness. They are (1) being thrown back, (2) being subdued, (3) being injured, and (4) losing a part of one's self, abandoning part or all of one's self. Of course the phenomenological analysis of the different types of pleasantness and unpleasantness should be continued. My few remarks will try only to indicate the direction of approach.

Generally we think of feeling and emotion as spontaneous acts. It may be important, especially for applied psychology and psychotherapeutics, to make clear that *what* a feeling, an attitude, an expressive movement or a gesture means can be changed by the situation-creating power of our words. Here I do not think of the intellectualization of feelings, exposed by Dugas.[18] It was Brice Parain[19] who explained that we create a situation by our speaking. When I say (or write), "It is now night," I choose to project an image of a world and at the same time a definite nuance of feeling. It is another feeling and another world from when I say, "It is becoming night." We are able to change, to strengthen, or to weaken our feelings by our language. We may further the genuineness, the truth, and the reality of our feelings or their falseness. Saying "I am hungry" serves to communicate to myself and to others that my trouble is being hungry. By this determination I call forth a situation with many consequences. My words – for example, saying to anyone, "I love you" – throw me into an adventure of affections and a series of situations and also serve to present me in a certain mode of existence. A word is not only a communication concerning a concrete attitude and situation; it also creates new situations, attitudes, and feelings.

It has been my intention to outline the phenomenological approach to the problem of feeling and emotion. I should like to end this short communication by trying to show that the signification of expressions – for example, a smile or a blush – can be understood from the phenomenological discovery of the essential sense of the situations in which they occur.

Dumas[20] concluded from the well-known experiments of Duchenne that "smiling is the easiest response of the facial muscles to moderate excitation." This conclusion is true, but it is not only a physiological moderate excitation that causes a smile. We must ask what is the essential nature and signification of a "moderate excitation" and find out what are the characteristics common to all situations which provoke a smile. Analyzing two situations that typically involve smiling, a friendly encounter and the making of gestures toward a person as if to tickle him, I came to the conclusion that a "moderate excitation" has a special meaning experienced in the intentional act. This meaning is not only that the excitation is small, but also

18. L. Dugas, "La logique des sentiments," in *Nouveau Traité de Psychologie*, VI, 1 (Paris: Alcan, 1937).
19. Brice Parain, *Recherche sur la nature et les fonctions du langage* (Paris: Gallimard, 1954).
20. Georges Dumas, *Traité de Psychologie*, III, 2 (Paris: Alcan, 1937), p. 344.

that it will be kept small, kept in check in regard to something else which is anticipated in the encounter. We experience *in* the slight joy, the momentary contentment, a greater joy in the future, without knowing exactly what we are anticipating. For this reason the "moderate excitation" which we reply to by a smile has an ambivalent structure. It is similar to the situation of timidity which elicits a smile so easily.

What does "the easiest response" mean? Dumas writes:

> It is merely choice ... concerning our cheek and eyelid muscles: we should smile differently if our facial muscles were attached in other relationships or moved differently, and if, by chance, the grimaces which occur in pain had been the easiest facial contractions, those would certainly have constituted the human smile.

When we conceive the body as contingent, the supposition of Dumas is plausible, but when we are sure that human (and animal) being is a psychophysical totality in the sense that there is an essential significant relation between the organization by which the individual exists in the world and the possibilities of the intentional acts which involve being in different situations, then is it inconceivable that the facial expression of pain could be the easiest?

Observation of the young child teaches us that smiling is easy and that crying requires an effort.

The smile wells forth from the unstable tranquil existence that already possesses a background of complacency. This comfort increases in the presence of the other person as a warmth of feeling that fills our existence and comes over us like a tidal wave. This occurs spontaneously, easily, and the irradiating, overwhelming complacency is in us just as a feeling of satisfaction may be in us or the vital cheerfulness after drinking alcoholic beverages. When smiling we do nothing but just smile – "the easiest response" to the moderate excitation of complacency.

The unstable tranquil existence, which is the condition for smiling, is not a sleepy, passive state of consciousness. It is a definite, easy attitude, the attitude of rest, of an active inactivity. We can observe it in every baby that lies awake. We observe it in ourselves in quiet sitting or walking. It is an attitude of confronting environment, which, as a visible world, supposes and composes our existence.

The paradox of smiling is that an activity signifies a relaxation of an active attitude of rest. The smile is in this way the expression of a threshold situation, of an impending outburst of joy, of the opening of a closed being, of a self-sufficient, immanent complacency, combined with a transcendental, anticipated cheerfulness. In the expression of the smile we observe the instability, the twinkling, and the sparking of joy as well as the stability, the durability, and the seclusion of tranquility. It is the ambiguous expression of an ambiguous act.

Already in the first smile of the child we observe the expression of a threshold situation, the first timid, sympathetic encounter. This smile is an answer of the human being in which it constitutes its own transcendental ego, its being-for-self. At the same time the child becomes confused with its own self, leaving the vital immanence and transcending the seclusion of tranquility, finding itself on the threshold of tender communication, called in to this situation by the mother, the matrix of love.[21]

What physiology means by "moderate excitation" and "the easiest response" gets its full significance in the phenomenological approach to human existence, feeling, emotions, attitudes, situations, and expressions.

When feeling and having emotion, man is physically in a definite way involved in the "magic" world he is projecting (Sartre). Always, and this is important to establish, he is *also confronted* with this projected world as with something objective, present, and perceived. This dual relationship is contained in the dynamic structure of feeling and emotion and is to be detected in many human expressions. I think we can find this dual relationship in blushing, the dilatation of the skin vessels in shame, the only *constant* outwardly appearing change. In this case, there *seems* to be no question of any relation to an intentional act and therefore of any signification. Still we are here faced with a phenomenon which is not related to a certain biological situation, but to a specific human mental state, which does not prevent blushing from being quite involuntary, without any reflection.

Blushing for shame, as has been known by scientists as far back as Darwin, takes place among all peoples. Blind persons blush; idiots, however, do not. Children blush more easily than adults; girls more easily than boys.

Shame, as well as shyness or timidity, elicits blushing under certain circumstances, *e.g.*, when blundering, when being praised excessively, when trespassing against the rules of convention, when being caught in wrongdoing, when accused falsely.

It is not possible to explain blushing physiologically. Even Dumas takes this view. He tries to explain blushing on psychological grounds and points out that it appears in those cases where normal reactions are inhibited, impeded, or falsified. In such cases we can observe the resultant nervous processes, sympathetic effects like sweating and vasodilatation. According to Dumas, the perspiration of a student being examined would be such a resultant phenomenon as well as the blushing of someone who is ashamed. This explanation does not seem to be very satisfactory. Its only advantage is that it draws our attention to the fact that man blushes when he finds himself at an "impasse," when he cannot find a way out and finds himself incapable of doing or saying anything.

In considering the various circumstances which may bring about a feeling of shame, it seems that the general chracteristic is the discovery, the unmasking, the exposure of our unworthiness.

A young girl entering a drawing room cannot possibly return every glance thrown in her direction. Such glances do not require any return. They convey a certain opinion, formed independently of her words or actions. The real intentional act on her part is the recognition of the indiscreet glance for what it is. It is therefore the dialectic relationship that is lacking, the interaction between the person entering and those already present. Inasmuch as the girl suffers – or experiences –

21. This is amply discussed in F.J.J. Buytendijk, *De eerste glimlach van het kind* (The First Smile of the Child) (Nijmegen: Dekker en van de Vegt, 1947).

their glances, she is not alone but in the others' presence. The others are in proximity to her only as her critical observers, and more specifically the observers of her bodily presence. The glances directed at her hands, her feet, her belt, the style of her coiffure, and her back indeed touch these parts of her outward appearance, but through these, they touch herself as she exists physically *for others* in these hands, this way of dressing her hair, etc. All parts of her body and clothes are suddenly felt as her own, but at the same time as insufficient to cover her physical being in the world. Not capable of parrying these glances by returning them, the person experiences them as penetrating, discovering, unmasking. Attitudes, movements, words, and even clothes lack the power to meet, to divert, to hold those glances. One is at a loss as to what to do, loses one's self-confidence and one's self-esteem. Shame is the feeling of existential *unworthiness* and blushing is the expression of this feeling. Sartre has in the chapter entitled "L'existence d'autrui" in his *Être et le néant* described the characteristics of shame. He concludes that shame is the apprehension of a unit of three dimensions: "I am ashamed of *myself* before *others*." The essence of pure shame is in his opinion "to be a thing."

But why exactly blushing and not something else? No matter how often a slight feeling of shame occurs, particularly on account of a clumsy remark, an offense against etiquette, etc., most people, even adults, will find it difficult to recall exactly what they experienced while blushing. Most of us are so strongly armed by the certainty of acting aright that we are able to respond in all situations without having the feeling of being virtually undressed. Only in our dreams is this different. When dreaming, the feeling of helplessness and unworthiness may come over us so strongly that we are deeply ashamed. We remember this very well after waking up. We don't know if we have blushed, but we do know that we felt oppressed and hot to the point of suffocation. This feeling is the vital, dynamic component, which necessarily accompanies the realization of existential unworthiness. Being ashamed is indeed not only an acceptance of suffering; it also contains an act of resistance, a revolt. This is explained by the fact that helplessness is the result of failing to do something which one should have been able to do and ought to have done. If we call this the ethical component of shame, the revolt will be its intentional correlate. Now we can also understand why we should feel oppressed to the point of suffocation when ashamed. It is the expression of an "effort," an unsuccessful effort, a superhuman effort to conquer an inadmissible and therefore unreasonable impotence. Once more I refer to dream experiences. The same feeling of oppression we experience when being deeply ashamed we feel during the well-known impotence when we dream that we are trying to run and cannot. So far as our dream consciousness can see, every condition for running is fulfilled in the situation constituted by the consciousness when dreaming. No cause to prevent it is experienced. And nevertheless it is impossible to run. An imperceptible, viscous mass seems to hold us back. We get hot and oppressed and may very well wake up wet with perspiration. In nightmares it is not an emotion that disturbs our inner existence; it is the *existence itself* that is changed. *Shame is a nightmare being lived while wide awake.*

Blushing is an outburst of oppression which finds its source inside us, in an unacceptable unworthiness, and which finds its outlet in our face, because that is where our self-being is really in the world.[22] The emotion from within overflows the cheeks in the form of a blush.

Now why is there no perspiration? This does occur occasionally, but as a rule it does not. There are fewer sweat glands in the cheeks than in the upper lip and the skin of the nose. Moreover, reddening is not a real reaction to warmth, but an outburst of effort in the revolutionary action of the individual. A physiological predisposition to blushing exists. We notice it only in people with a highly reactive vascular system whose color also changes for other reasons. That is why women and girls blush more than men and boys. But this is not the only reason. The chief cause of blushing lies in the situation as projected by feeling, and that is, under similar outward circumstances, different in men and women, in the young girl and the adolescent boy. To be discovered, undressed, seen through whether literally or figuratively, has an entirely different meaning for both sexes, as men and women experience their own bodily existence and their being-in-the-world in such different ways. As explained elsewhere, sexual life does not determine the nature of men and women, but their mutual existential relation, and consequently the way they experience sex, results from the essential difference between the sexes.

Woman's inclination toward taking care of people and things expresses itself even in the young girl in a form of self-grooming through which she experiences her bodily existence as her own much more strongly than boys do. She is, and feels herself to be, in contact with others through her body, but only through the fact that her clothing is seen *before* her and at the same time *for* her. It is this clothing that gets such full attention during a girl's adolescence. It is during this period that girls discover the dual function of clothes: to cover and to uncover. Both functions require a personal relation to clothes. In the period during which this relation is realized, self-esteem is most vulnerable in the matter of clothing. This is the period when the slightest occasion will bring about blushing and when the sensation of shame on being uncovered is greatest. A very slight stimulus – a word, a question, a glance – suffices to make one feel naked, and therefore helpless, inasmuch as one is stripped of all means of covering and uncovering on which well-balanced social relations and self-esteem are based. Like a flame, *hot* and consuming, the feeling of shame wells up from within into the face, the part of the body through which we are in the world.

The vegetative effects of emotion may certainly often be meaningless, unspecific

22. We experience our self-being as present *in* the face. There is another explanation for this besides the social function of the face, besides its being the uncovered part of the body which we turn to our fellow men. Independent of any relation to others, our transcendental "ego" is present in our face, or rather, right behind it, so that the ego "looks out through" the eyes. If we ask a person to indicate the plane separating things in front from those behind him he will indicate a plane going through the ears; the separation between things above and below him will go through the eyes. Feelings connected with human (ethical) relations express themselves exclusively in the face.

"Envy, shame, remorse, desire are limited to the facial expression and make the face the real mirror, the sounding-board of feeling and emotion, specifically through growing pale or red." (Plessner, *Laughing and Crying*, pp. 44-45.

side phenomena, irradiations of stimuli in the field of the sympathetic nervous system. In animals they may have a certain type, which then may become purposeful characteristics of the emotional reaction. In man, however, they come to a phenomenon expressing the essential nature of the feeling and emotion. The most striking example of this is blushing for shame.

To understand the value of the phenomenological approach to the problem of feelings and emotions, we have to establish the value of description in general and distinguish between the introspective and the phenomenological methods. Introspection is always an inspection of subjective experience. I describe this kind of experience exactly in the same way as any observer gives a report on his sensory simultaneous and successive perceptions. In both cases we observe facts, their positions in time, and in a real or imaginary space. When I should like to describe the mutual relations of these facts, I must have a categorial system of references, for example, the physical concept of causality or the metaphoric and often mythological notions used in physiology and psychology. The phenomenological description, however, is directed immediately toward experienced phenomena, as well as toward the different acts – the acts of perception, remembering, thinking, feeling, etc. All these phenomena are taken exactly as they belong to the phenomenal world. This world is given to us directly and shows itself in itself, in its meaning. The question whether the phenomenon is real or not is disregarded. The only interest of the phenomenologist is directed toward the *essential* structure, the intrinsic connections, and the self-revelation of the significance in the full context of the phenomenal field. So the question, "*What* is happiness or shame?" is the same question as the one put in common life: "What is a chair or a box?" or "What is red?" In both cases we are not satisfied with the indications of the conditions necessary to produce these phenomena, neither with the description of the material of which they are made nor with the result of an investigation into the causal relations to other things. I want to know exactly what kind of thing a chair or a box is, what it signifies in the world, where chairs and boxes are experienced, and what is the inner *essential* structure by which these significations are clarified. Just as in common life, phenomenology wants an evident insight in the sense of experience, for example, in feelings and emotions.

The continued research on feelings and emotions as stimulated by the phenomenological approach will require exact and repeated observation of the phenomena in the above-mentioned sense of this notion. This observation or contemplation must be performed with the aim of penetrating to the essential sense as it is presented in the intentional act itself. Of course our experience is never complete. Everyone who has read the *Farbenlehre* of Goethe will have learned how "rich" the content of a simple perception of color is. The word "red" or "violet" is neither only a sign to indicate something nor only an abstract notion, but we use these words also to express our direct experience of these colors. In daily life as well as in science a word generally functions like a chain to get hold of a dog, to have the dog near us, and to do with it what we like. The phenomenologist, however, is not interested in the chain but in the dog itself. This is what Valéry meant when he wrote: "Every word is a bottomless pit."

In this way we understand that the development of the phenomenologically orientated psychology depends on the dialectic relations between exemplary contemplation and the verbal expressions in living language developed out of unreflected intuitive views, signifying the essential sense. Perhaps one would assume from this that the phenomenology of feelings and emotions may be better furthered by artists than by psychologists. First of all I should like to reply that art has proved its high value for psychology; second, that psychology requires the most thorough description of experience; and third, that applied psychology has proved the importance of the phenomenological method along with the analytic, functional, statistical viewpoints. It is impossible to know what an abnormal feeling is without an insight into the essential significance of normal feeling in the context of human existence. It is impossible to understand or even to describe sex relations when we do not know the essence and meaning of sexuality, bodily existence, femininity, or masculinity. Pedagogics requires understanding the full content of sense of the phenomenon of play as well as of the mode of existence of the adolescent.

We are at the beginning of the phenomenological approach to all these matters and to the feelings and emotions in particular, but it is desirable for us to be aware of the richness of phenomenological analysis gathered in ancient and recent psychological literature, for example, in the publications of *Gestalt* psychologists.

An important starting point for further research is, among other things, the statement of William James: "We ought to say a feeling of *and*, a feeling of *if*, a feeling of *but*, and a feeling of *by*, quite as readily as we say a feeling of blue or a feeling of cold." And he adds: "Yet we do not: so inveterate has our habit become of recognizing the existence of the substantive parts alone, that language almost refuses to lend itself to any other use."[23]

The progress of phenomenological analysis of feeling and emotion depends on the discovery of the invariance which occurs during the continual experience concerning a feeling in various situations in relation to the modes of existence of normal and also of neurotic individuals, of men, women, children, primitive people, and animals.

The implicit conception of human reality and of the characteristics of human existence must become explicit during such a research through the evidence of insight into the signification of every intentional act of a concrete mode of feeling.

Repeatedly the question has been raised as to what is the guarantee that the phenomenological statement is true and not a pure subjective opinion. The answer to this is simple. The evidence of a statement about the essence and sense of a phenomenon can be proved only by the affirmation of everyone who, without prejudice, directs himself to the phenomenon in question. It is possible that some observer will find new, more detailed, more specific essential characteristics. Scheler, who has enriched us with so many phenomenological analyses – for example, about repentance, resentment, and sympathy – once said that the phenomenological ap-

23. William James, *Principles of Psychology*, 2 vols. (New York: Holt and Comp., 1918), vol. I, pp. 245-246.

proach offers the advantage of a continuous progress of insight after every return to the "thing-itself in its thisness." Such progress is the aim of science.

7

Eidetic of the Experience of Termination

STEPHAN STRASSER

1. FEELING OF RESULT

However diverse the notions may be which we form of "happiness-as-such", in one respect we agree essentially: we know that we do not possess it from the beginning. One "makes," "seeks," "builds" his happiness. In all these expressions we find the notion that happiness must be the result of an act, an activity, a phase of life, the result of living. The experience of happiness thus has nothing immediately to do with the inchoative upsurge and being-underway in a certain direction. It is rather a kind of experience of termination. It is connected with the terminating phase of experience in a manner to be still more carefully determined.

Now, this end-phase is particularly significant from a phenomenological point of view. It is characterized, among other things, by the fact that straining ceases; a certain relaxation is noticeable; a small pause indicates the end of a rhythmic *Gestalt*. But that is not all. If the result of an act is positive, then for a moment it fills us completely. It is, as we have already indicated, as if we stood before something infinitely positive, and the contemplation of its value, splendor and significance totally filled us. But it does not last long. This "infinity" does not bring the stream of our activities to a standstill. We are very soon able to reflect and to distinguish. We fit within our world-picture what we have discovered; we compare the additional gain with that which we already possess; we try to assess the range of our success. Then we want to be aware of the fact that the elevation and enrichment of our being is not merely quite relative. What appears to us the Whole turns out to be a part; "the" uncovered Truth shrinks to "a" truth; Infinity changes into finitude. An inevitable disillusionment thus sets in. The restful feeling of result changes into one that is agitating, *quietatio* becomes *ira.*[1] Dissatisfaction with the unsatisfying result provokes us to new deeds of which we hope that they would be more successful, more real, more intelligent than those that went before. And thus the dialectic of

* "Eidetic of the Experience of Termination" appeared in *Phenomenology of Feeling* by Stephan Strasser. The book was published by Duquesne University Press in 1977. The passage here reprinted by permission of the Publisher is found on pp. 327-348.

1. Cf. Stephan Strasser, *Phenomenology of Feeling*, trans. Robert E. Wood (Pittsburgh: Duquesne University Press, 1977), pp. 226ff.

the incomplete completion enters into a new phase. It continues – always on a higher level.[2]

It is certain that the experience of happiness has something in common with that of lived termination as we have just described it. But is any positive feeling of result thus comparable to an experience of happiness? Such an identification would be equivalent to a simplification and idealization of our set of problems that we cannot allow. Let us consider a bit the everyday experiences. No one has been able to observe that a factory worker at the conveyer belt breaks into jubilation over the result of his work, or that an accountant is made happy over a correct calculation. To be sure, it is true that every theoretical and practical act which attains its goal affords occasion for a certain satisfaction. But this receives its experiential *Gestalt* in many cases in the form of a slight relaxation that is experienced as agreeable. Expressions like "We've finished that again!" show this slight satisfaction. We could perhaps with William Stern, call the thymic echo which this situation arouses "feeling of result."[3] Yet in this way, the experience is not necessarily bound up with a delightful, fascinating infinity. The happy experience is thus completely different than the modest awareness of attaining a positive result. Between both phenomena there is a connection; but it is not easy to say wherein it consists. Certain "simple" solutions to problems which intrude do not stand up to a critical testing.

At first sight, it appears as if any awareness of result, even the most superficial and fleeting, contains a bit of happiness. Many of these "*petites perceptions de bonheur*" (as one might freely turn a phrase from Leibniz) must then lead to an authentic experience of happiness. It would follow from this that man, because of his goal-conscious work or his continual fulfillment of duty, must become always more contented, and finally, perfectly happy. This atomistic conception of the essence of happiness, for which an elemental-psychological prejudice ultimately furnishes the basics, answers very badly to the phenomenological reality. For example, consider the case of our accountant. After a thousand additions, he will be scarcely happier than after the first. Perhaps he will even feel a bit unhappy because of his intellectual exhaustion.

One might perhaps come to a better theoretical position if one completed and built on the phenomenology of the act. It is clear that the analysis of an isolated act does not lead to the mastery of our problem. For that it is necessary to include in our treatment human activity in its totality. Now this activity, as we know, constitutes a complicated, but yet ever-ordered and hierarchically stratified totality.[4] Let us begin once again with the example of our accountant's employment. The arithmetical operations which he has to perform, day-in and day-out, scarcely interest him. He works primarily because of the salary that he receives at the end of

2. For this analysis cf. L. Vander Kerken, *Het menselijk geluk* (Amsterdam: Noordnederlandsche Boekhandel, 1952).
3. Cf. his "Vierfelder-Schema," in *Allgemeine Psychologie auf personalistischer Grundlage* (The Hague: Nijhoff, 1950), p. 736.
4. Cf. Stephan Strasser, *op. cit.*, pp. 240ff.

the month. But this sum of money is a means and not a goal for our man. He saves for a year and a day for a trip. This trip – so he hopes – will provide him with a happy experience. The levels of human activities would, accordingly, be ordered by reason of their finality. Activity (a) (in this case his professional activity) has as its goal making activity (b) (saving) possible, while (b) seems meaningful only in reference to act (c) (the planned trip).

One could thus formulate the general thesis that human activities correspond to diverse levels, that these are ordered hierarchically according to well-established basic fixed principles, and only the success of the acts of the highest level ought to be treated as producing happiness. Only they would be immediately meaningful and suited to the goal. In order to describe the experience of happiness, one has thus only to take into account such acts as bring about a positive termination in human life or at least in a phase of this life. We could characterize as "happiness" only the definitely favorable turns in the course of one's life.

This theory corresponds completely with the popular conception of the nature of happiness. In everyday life an appointment, a promotion, an exciting result (perhaps winning a lottery), a marriage, the birth of a child, homecoming from confinement and so forth, are treated as happiness-producing events. Still, the scientific as well as the popular approach suffers from the malady of objectivism. Tacitly one accepts the presence of a sociologically "given" which governs the hierarchical stratification of human activities once and for all. One also finds it evident that certain objectively describable events "cause" happiness. But do things really happen this way? Is it not man himself who, on the basis of his evaluations, creates and supports the hierarchical ordering of his activity? And are there events which can be characterized as causes – in the natural-scientific sense – of happiness? Have there not been weddings and baptisms which make a heartbreakingly sad impression? Cannot the return from confinement also turn out to be tragic, etc.?

As far as the problem of stratification is concerned, a comparison of two typical cases would be extremely instructive. Let us consider once again our worker at the conveyer-belt. His activity is scarcely satisfying to him. It is quite apparent that he wants to perform it solely for material profit. Perhaps it is different with a graphic artist (for example, an artist of the type of a Vincent van Gogh). In him the ultimate stratification of activities shows the reverse picture. Perhaps it is precisely the daily struggle with the reluctant materials that he finds fascinating, while the monetary profit only interests him insofar as it would allow him to continue the struggle. Furthermore, we also know from our own experience that the appreciation of our own activities can basically change. What at first promised us satisfaction and joy turned out to be an empty routine, while a new ideal appeared on the horizon of our life which, in an instant, attained the rank of a highest purpose. Accordingly, *there is no psychological or sociological law which makes it possible for us to determine once and for all the characteristics of the happiness-producing act or level of acts.*

According to what has been said, it is clear that the value-structure and the ultimate ordering of the various activities ultimately depend upon the way in which man deals with the world. Accordingly, *the question of happiness is only to be answered from the point of view of the actual basic comportment* which the man in question takes up in regard to the *mysterium* of his being. In order to know wherein his happiness consists, one must know his existential project. The experience of happiness always occurs before the transcendental horizon, even when its cause appears to be a banal event. In that respect, it is basically distinguished from the agreeable, pleasurable or satisfying impression which the success of an act, however formed, leaves in us.

Hence the feeling of result and the experience of happiness are phenomena which belong to two very different levels. The former can basically be exhibited also in performances limited to the environment; the latter displays a transcending character. Between the positively terminating feeling and the experience of happiness lie various intermediary levels which cannot be skipped with impunity. Thus it will be our first task to present a description, however summary it might be, of a few phenomena which make the transition from simple awareness of result to the experience of happiness appear plausible.

2. ENJOYMENT (*GENUSS*)

If we wish to determine the essence of enjoyment more carefully, we must first of all encounter the deeply rooted prejudice according to which pleasure and enjoyment are essentially the same, and the latter is distinguished from the former merely through greater intensity and the complexity of its object. The fact that enjoyment is only possible in the region of terminating experience points already to an essential distinction between both modes of experience. What is enjoyed is, in other words, always what has been attained, achieved, acquired. In the case of pleasure it is different. Pleasure already accompanies performance in the phase of being-underway, and indeed, as we know, in a double form: as the pleasure of functioning and as prospective pleasure in the result.[5] Pleasure can thus operate in elementary performances as a guide. But there can be no talk of a "guidance by enjoyment," since the one who enjoys is already at the goal.

But from this one ought not conclude that enjoyment is identical with pleasure in the result. The proof of the untenability of this view is provided by the fact that ambivalent results can also be enjoyed. The pleasing character of the experiences of scratching and itching or of the good taste of "piquant" (i.e. literally, "biting") foods indeed rests upon precisely a peculiar mixture of pleasure and displeasure. On a higher level we find analogous phenomena. Thus, for example, the "ambiguous" or "gruesome" character of a story or the "blissful anxiety" in carrying out a forbidden activity can become consciously enjoyed. The Alpinist enjoys climbing a mountain not in spite of, but because of the dangerous position, etc.

All of this requires careful analysis. Still it suffices to establish what Vander

5. Cf. *ibid.*, pp. 215ff.

Kerken has already stated: that the pleasurable – and we add, also the unpleasurable – is subsumed by enjoyment into a higher unity of awareness.[6]

But this negative delineation in no way suffices to characterize the nature of enjoyment. One would probably be inclined to accept that talk of enjoyment is in place when one who desires it possesses a concrete good which gives rise to pleasant sensory perceptions.[7] With that we come closer to the situation. Enjoyment is essentially enjoyment of something; to it belongs further the "possession" of the intentional object. But what does the word possess mean in this connection? Talk of *possessio* is figurative; we must first inquire into its proper meaning. We must also investigate whence it happens that whole categories of goods can be desired but not enjoyed, namely all "goods of the way": tools, instruments, medicaments, etc.

Furthermore, the notion that all enjoyment is of a sensory nature urgently needs checking. Also in daily life we sometimes affirm that we "enjoyed" the elegance of a mathematical demonstration, the cleverness of a dialogue, or the facility of a technical performance. It is true that, in the latter cases, enjoyment is founded in perception; yet, according to its essence it transcends the sphere of mere perceiving. One does some violence to the facts if he equates "enjoyment" with "sensory enjoyment."

Finally, the question must also be raised whether, in the case of humans, the so-called sensory enjoyments really are what certain all-too-abstract thinkers consider them to be. On this question it is instructive to consult the artists who possess the gift of reproducing the *essentia* of the concrete. For example, take the description of the enjoyment of smoking tobacco mixed with belladonna which Felix Timmermans places in the mouth of painter Adriaan Brouwer: "Oh, to suck the bewitching spirit of tobacco from the claypipe, to blow smoke rings, to gaze at them and to becloud oneself with a trail of smoke! What a glorious life! It is as if the world with all its misery, its ardent pulse beat and its egoism ... sank into the fog. Nothing troubles you any more, you no longer feel the stone press upon your heart, and you yourself sink away, you disappear into a deep, sweet stupor...."[8] Now the enjoyment of the smoker, as Timmermans describes it, is doubtless the prototype of a sensory *delectatio*. Still it will be difficult to say in this case wherein the merely sensory lies. The opinions, thoughts, and views which Timmermans so appropriately

6. L. Vander Kerken, *Het menselijk geluk*, pp. 126-127. Philipp Lersch expresses himself similarly: according to him, every enjoyment "contains" pleasure (*Aufbau der Person* (Munich: J.A. Barth, 1952), p. 185). On the problem of ambivalent experience, cf. among others: E. Bleuler (*Dementia Praecox oder Gruppe der Schizophrenien* (Leipzig: Deuticke, 1911), pp. 305ff. and 395ff.) and Karl Jaspers (*General Psychopathology* (Chicago: University of Chicago Press, 1963), p. 342). On the problem of ambivalent experience of Transcendence, cf. Rudolf Otto, *The Idea of the Holy: An Inquiry into the Non-Rational Factor in the Idea of the Divine and Its Relation to the Rational*, trans. W. Harvey (Cambridge: Oxford University Press, 1950).
7. In this sense, one could relate Victor Cathrein's descriptions of *hedone* to enjoyment, but not to that which the psychologist is wont to term "pleasure." Cf. "Lust und Freude, ihr Wesen und ihr sittlicher Charakter mit besonderer Berücksichtigung der Lehre des Aristoteles," in *Philosophie und ihre Grenzwissenschaften*, vol. III (Innsbruck: Rauch, 1931), Heft 6, pp. 12ff.
8. Felix Timmermans, *Adriaan Brouwer* (Amsterdam: P.N. van Kampen, 1947), p. 27. In the translation we had to give up rendering the shadings of dialect in the original text.

renders, are indeed no abstract meditations upon enjoyment; they rather belong together with the enjoyment, indeed they are what makes the enjoyment to be enjoyment. Hence one can well say that sensations of pleasure constitute the necessary condition for sensory enjoyment; but it is wrong to suppose that they represent the sufficient condition.[9]

What is meant by talk of sensuality is perhaps something else entirely. It is essential for sensory enjoyment that its object can only be something concrete and individual. But that also holds true for enjoyment in the broadest sense of the term. It is impossible for us to enjoy the ingeniousness of mathematical proofs-in-general. What we enjoy is the way in which *this* mathematician solves *this* problem *now*, before our spiritual eyes. What will be enjoyed is the unrepeatable beauty of a definite moment – hence the *carpe diem* of the ancient artists of life, the inimitable elegance of an individual style of thinking or the masterfulness of a concrete aesthetic performance. What will be enjoyed is the bouquet of *this* wine, the aroma of *this* tobacco plant, the good taste of *this* food, *here and now*. Characteristically, we do not call immoderate eaters and drunkards "enjoyers," and just as little do we so term the advocates of the Don Juan type. For *the art of delaying* belongs to enjoyment. One who draws from the moment, the situation, the object, all that it has to offer of splendor, graciousness, spiciness, or cajolery, understands enjoyment.

If the one who enjoys is in many respects the opposite of the adventurer, the man of action and the passionate personality, the basis of this lies in his receptivity without activity. Receptiveness, preparedness for assimilation, and sensitivity attuned to the finest nuances belong to enjoyment. To enjoy consists in consciously letting flow into oneself what is proper to the object as far as gratification is concerned. "All enjoyment is assimilation.... In its essential passivity and receptivity, enjoyment is the opposite of willing and striving," Moritz Geiger maintains in a perceptive investigation.[10] Geiger basically provides Aristotelian wisdom in modern garb: "Activity is not only given in the form of movement, but also in the form of lack of movement, and enjoyment consists rather in rest than in movement," says the Stagirite.[11] Also in the Scholastics, *delectatio* is brought into relation with the cessation of the act, with *quietatio.*[12]

The satisfaction of all seeking and pursuing, restful rejoicing in the object: that is what is literally signified by the image of possession. In social life the untroubled and secure "having" of a concrete good is possible only in the form of *possessio*. That one who enjoys "possesses" the good thus means that he has it completely in his power, that he has to make no further efforts, that there is no more danger to

9. Stephan Strasser, *op. cit.*, pp. 167ff.
10. "Beiträge zur Phänomenologie des ästhetischen Genusses," in *Jahrbuch für Philosophie und phänomenologische Forschung*, vol. I, Halle, 1913, p. 605.
11. "... for actuality (*energeia*) exists not only in motion (*kineseos*) but also in something motionless (*akinesias*), and pleasure (*hedone*) depends more on rest (*eremiai*) than on motion." *Nicomachean Ethics*, VII, 15 1154b26-28. (In the original the Greek text was quoted in this note; I have translated this passage "literally" and indicated the Greek words which are important in this context in transliterated form. Ed.).
12. Stephan Strasser, *op. cit.*, p. 212.

fear, no more envy to control. One who enjoys can appreciate what he has attained in complete freedom and rest.

But this does not imply that only the solitary one, only the *individuum* or only the *solus ipse* can enjoy. The opposite is true. The awareness that others, friends, experience the same thing, can heighten one's own enjoyment. Yet those who enjoy with you must be persons who stand closest to the I and do not belong to a random group or mass. The tendency toward exclusivity dwells in enjoyment. In opposition to emotion and passion which operate infectiously and can travel in company with masses, enjoyment leads to "*solitude morale*" as Georges Dumas noted.[13] Enjoyment is something intimate. One possesses thereby something that one does not wish to share with everyone. That is true for the secret society as well as for the bond of friendship and the family circle. Goethe gave expression to this truth in the famous verses:

Selig, wer sich vor der Welt *ohne Hass verschliesst,* *Einen Freund am Busen hält* *Und mit dem geniesst,*	Bless'd the man who hateless hides From the world's employ, By a bosom friend abides With him to enjoy,
Was von Menschen nicht gewusst, *Oder nicht bedacht,* *Durch das Labyrinth der Brust* *Wandelt in der Nacht.*[14]	That which, though not manifest, Or which thoughtless slight, Through the Lab'rinth of the breast Wanders in the night.

Goethe's words indicate a further essential feature of enjoyment: it is always conscious. Indeed, one can say immediately that the intensity of enjoyment grows in proportion to the awareness of enjoyable assimilation. Thus we can explain why a wine connoisseur appreciates features of a glass of wine which completely escape the non-expert, or why only the mathematically educated can appreciate a geometrical proof, and so forth. Refined knowledge of the object, acquaintance with the relations of rank in a determinate region of value, and receptivity for nuances doubtless heighten the capacity for enjoyment.

Wherein then does enjoyment differ from the experience of happiness? The distinction is simply that enjoyment is never a transcending experience. Whoever enjoys is receptively related to a good or to a region of goods. The smoker knows that his enjoyment has for its object the aroma of a plant; the wine connoisseur concentrates upon his taste impressions; the aesthete upon the fine nuances of the work of art. *Carpe diem!* means basically: "Use the day, because it is a finite, fleeting good." This is the opposite of basic transcending comportment. In the erotic realm, the opposition is particularly clear. One who "knows women" surrenders himself consciously to sexual enjoyment. For the passionate lover, on the contrary, sex becomes a vehicle for all the heights and depths of encompassing experience.

13. Georges Dumas, "La douleur et le plaisir," in *Nouveau Traité de Psychologie*, 9 vols., (Paris: Alcan, 1932ff.), vol. II, p. 290.

14. Johann Wolfgang Goethe, "An den Mond," in *Sämtliche Werke*, ed. Karle Goedeke, 15 vols. (Stuttgart: Cotta, 1871-1872), vol. I, p. 72.

He is primarily related to the human Thou, while at the same time he intends to view the fullness of value through the medium of this Thou. He is, in a certain sense, diverted from actuality ("*distrait du réel*," Renée Dejean would say). His happiness lies in a transcending experience which is not comparable with a mode of behavior in which he merely enjoys the beloved.

3. JOY

a. The Essence of Joy in the Universal Sense

That the phenomenon of joy is *toto caelo* different than that of enjoyment appears to us clear at once. Even in prescientific life we use both notions in markedly different ways. But it is not easy to give a more exact account of the difference. Perhaps a more exact analysis of terminating experience can be useful to us in this regard.

It is evident that, in the phase of completion, two moments are to be distinguished: the moment where the subject attains certainty that he will reach the sought-for good, and, on the other hand, the state of untroubled and painless possession of the good itself. If the latter situation is given in experience, then we have before us, as was already shown, the phenomenon of enjoyment. *Delectatio* consists in an immediate cognitive *possessio* of a concrete good. But *possessio is not the* same as *adeptio*; possession should not be confused with taking-into-possession, attaining, mastering. Genetically, the latter precedes the former. The question is whether the attaining of the desired good does not give rise to a certain stirring of the heart.

Doubtless not every *adeptio* produces joy. The real act of mastery can also succeed in a manner that is completely dry, matter of fact, automatic. For example, if we buy a ticket, we do not break out in jubilation. We experience joy apparently only when we have previously persisted – in the phase of being-underway – in uncertainty or in anxious tension with respect to the result of our activity. In a routine action, no uncertainty regarding the result holds sway; thus it is also performed, for the most part, in a manner that is completely "joyless."

Hence it follows that *joy in its most general sense is the felt affirmation of attaining possession after a previous uncertainty*. As such, in the course of the act, it stands precisely at the boundary between the directed and the terminating experience. It has something in common with both phases. Joy gives to the phase of being-underway the mark of tension; it attains its highpoint in a moment of mastery. (This is manifest also in physiological symptoms such as rising blood-pressure, heightened tonus of the musculature, rapid pulse, etc.) On the other hand, joy already displays the clearly pleasure-toned character of termination. It is the conscious beginning of the heightening of one's own being; it corresponds to the *attingere* and *pervenire* of the Scholastics.

Furthermore, joy is distinguished from enjoyment in the fact that its immediate object does not necessarily have to be a good-in-itself. The discovery of a way, a means, an instrument, a method can also evoke joy. For example, one who stands before the locked door of his house would suddenly be overjoyed if he finds in the lining of his overcoat the key he had so long searched for. At the same time, one must, of course, note that the possession of the key is equivalent to free entry into his home. He attains complete certitude of attaining the goal that is good-in-itself. Where this is not the case, the possession of a means, an instrument or a method gives rise to hope, but not to joy. For example, one who goes hunting with a gun hopes that he will get something; he can rejoice if he is first certain he has bagged the wild animal.

Certitude is thus a precondition of joy. This certitude need not necessarily rest upon rational insight; it can also be caused by a prepredicative grasp of a structure in the environment. The dog who notices that his master has the leash in hand is completely certain that he will now be allowed to accompany him out of doors. He therefore shows his joy through barking and through expansive movements. Naturally, his certitude rests upon "conditioning" and not upon insight. But the usual sequence of events affords the animal a subjectively perfect assurance of the fact that something good-in-itself is at hand for him.

b. Joy as Opening and Exultation

On the other hand, we do not find the joyful emotion in the animal, but only in man. Only in man's case does joy lead him to an inner being-apprehended. Only man can become so overpowered by joy that he no longer graps the condition of things. Only he is able to shed tears of joy.[15] He experiences that which brings joy as a gift which removes him from the everyday. The bodily bearing of rising becomes for him a jubilant *sursum corda*; to the external movements of self-opening corresponds a being-opened of the heart. He is able to translate his inner release into the language of kindness and affection toward other men.

It would doubtless be instructive to pursue the question of how the great poets of world literature have given linguistic form to joy. Here we will content ourselves with two very different examples. The emotional character of joy appears nowhere more clearly than in the passionate hymns of the psalmists and prophets. For example, one thinks of the following song of rejoicing from Isaiah:

> The desert and the parched land will exult; the steppe will rejoice and bloom. They will bloom with abundant flowers, and rejoice with joyful song. The glory of Lebanon will be given to them, the splendor of Carmel and Sharon; they will see the glory of the Lord, the splendor of our God.
>
> Strengthen the hands that are feeble, make firm the knees that are weak, say to those whose hearts are frightened: Be strong, fear not! Here is your God, he comes with vindication; with divine recompense he comes to save you. (Confraternity-Douay version)[16]

15. Cf. Helmuth Plessner, *Lachen und Weinen* (Bern: Francke Verlag, 1961), pp. 179ff.
16. *Isaiah*, 35, 1-4.

The emotional character of joy appears here in an especially clear manner. The rapture, the delight, the ecstasy which overtake the prophets in the vision of future glory take the most palpable forms. It is thus characteristic that, without exception, the pictures of Isaiah have the moment of *adeptio* for their object. What is described, in ever-new visions, is completion; but note: not the most complete completion, but *the self-completing completion*. The breaking through of the blossoms from the parched plants of the steppe point to this; the appearance of the Messiah among the nations; the moment when they will catch sight of it. It is not the glory of the Lord that forms the theme of this song, but the moment when man will become aware of this glory; not the complete redemption, but the first becoming-manifest of the work of redemption is glorified. In our terminology, we can say that Isaiah described the turning-point when the painful and disheartening route changes into the exulting certitude of having attained the *terminus viae*. Isaiah's song is thus apparently a complete expression of joy, since he describes in monumental images this passage from doubt and uncertainty to jubilant certainty.

We find a completely different stylistic medium for evoking the disposition of joy in the Romantics: it is that of the poetic journey. On the imaginary trip of the poet every conceivable beautiful thing is explored, "wandered-over" and – in the literal sense – "journeyed-over" (*erfahren*). For example, one thinks of the following poetic journey of Joseph von Eichendorff:

Ich weiss nicht, was das sagen will! *Kaum tret ich von der Schwelle still,* *Gleich schwingt sich eine Lerche auf* *Und jubiliert durchs Blau vorauf.*	What it might mean I do not know! Over the threshold I step and lo! Forthwith a lark aloft does fly Up in the blue with joy on high.
Das Gras ringsum, die Blumen gar *Stehn mit Juwelen und Perl'n im Haar,*	Round me the grass, the blooms so fair Stand there with jew'ls and pearls in their hair,
Die schlanken Pappeln, Busch und Saat *Verneigen sich im grössten Staat.*	Slender the poplars, the bush, the crops Slowly bend down their stately tops.
Als Bot' voraus das Bächlein eilt, *Und wo der Wind die Wipfel teilt* *Die Au verstohlen nach mir schaut* *Als wär' sie meine liebe Braut.* . . . .[17]	Messenger-like, the brook does flow, Crown of the trees the wind parts so, And, Oh! so secret looks at me As if it were my bride to be.

Here not only is the festival tone of the prophets lacking, but different feelings are given expression. Whoever uses Scholastic terminology would say that in the Messianic songs of joy *exultatio* holds sway, while in the travel-songs of the Romantics it is *laetitia* and *jucunditas*. The poems have only this one thing in common, that in both a series of *adeptiones* are described.[18] What Eichendorff celebrates again and

17. Joseph von Eichendorff, "Wandernder Dichter," in *Werke*, 4 vols., ed. Ludwig Krähe (Berlin: Bong, n.d.), vol. I, p. 40. However, note that *pasture* (*die Aue*) is the subject of the last lines. (Ed.)

18. "To travel means to set out for *discovery*, for inquiry into the unknown for the experience of what is suspected," notes Vander Kerken in *Het menselijk geluk*, p. 195.

again is the moment when the traveller sets out, when the song of the lark reaches his ear, when he becomes aware of the charm of the landscape, or villagers, castles, etc. Here one can likewise speak of an incipient fulfillment and completion. Still, it does not give rise to emotional rapture, but is experienced in the mode of joyful elevation. The joy of the travelling poet Eichendorff is closely connected with serenity.

c. Serenity and Serenitas Animi

The phenomenon of serenity has already been described by Philipp Lersch in so excellent a manner that we have only a little to contribute to it.[19] Lersch correctly emphasizes that serenity – as distinguished from joy – rests upon a disposition. Characteristic of this disposedness is the feeling of elevation, of being free from the "burden of existence" and the inclination to remain in the present without prospective concern. The simplest things of everyday life are thus greeted with warmth and gratitude; the fullness of existence is received as an ever-renewed gift.

Now, it can be the case that we have nothing else to see in the serene basic disposition but the expression of a harmoniously vigorous vitality (as with a healthy child) or a balanced relation to the environment and the world (happy family life, satisfying occupation, gratifying joy in creating, etc.). Thus it has already been said that serenity – at least that of the mature man – rests upon something lived and experienced, even when this "something" is not explicitly conscious. Certain experiences and complexes of experience are sedimented in the thymic region and constitute the basis in the heart which sets in motion the disheartened, irritated and depressed, as well as the serene dispositions.

Thus one should first think of serenity in terms of experiences of inner satisfaction, harmonious interpersonal existence and beautiful self-unfolding. Indeed, the possession of such a reserve of the inner sense of joy clearly appears to be a precondition for the appearance of the serene disposition. This reserve can be supplied from experiences of the environment and of the world, but also from transcending experiences. The awareness of the "presence of God," for example, can, in the highest measure, be the occasion of that elevation, freedom from care, warmth and gratitude of which Lersch speaks.

Again, it is an Old Testament singer who has provided classical expression to the latter disposition. Consider the famous psalm:

> The Lord is my shepherd, I shall not want.
> In verdant pastures he gives me repose;
> Beside restful waters he leads me; he refreshes my soul.
> He guides me in right paths for his name's sake.
> Even though I walk in the dark valley I fear no evil;
> For you are at my side with your rod and staff that give me courage.[20]

19. Philipp Lersch, *Aufbau der Person*, p. 252.
20. Psalm 23, 1-4.

It is clear that here a situation is described that is not so exciting an occurrence as in the song of Isaiah. Serenity is a basic disposition of life and not a moment in the course of experience. Of course, one must not overlook the fact that that which the psalmist expresses represents a lack of care of a higher order. It has nothing to do with the unburdened character of the child or of the one who is happily enjoying himself, but originates from a transcending confidence, composure and fearlessness. The dangers and terrors of existence are indeed seen and taken with complete seriousness, but an inner elevation and certitude makes it possible for man to encounter them with composure. It is not a question here of being free from the burden of existence, but of the peace of those hidden in God. In order to characterize this serene and composed disposition which grows from the elevating awareness of the presence of God, one could speak of a *serenitas animi*.

4. THE EXPERIENCE OF HAPPINESS CONSIDERED IN ITS UNIVERSAL ESSENCE

The experience of happiness is not distinguished by degree but by essence from the feeling of result, from enjoyment, from joy and from serenity. But this difference is not heterogeneity. Rather, something common inhabits all the phenomena mentioned. A comparison of the experiences of happiness with the related stirrings of the heart can thus contribute to the clarification of this situation.

The experience of happiness is first of all an experience of termination. In this it is related to the *feeling of result*. But that which is experienced as happiness is an exceptional mode of fulfillment and completion. Thus the experience of happiness is distinguished from the feeling of result which, for the most part, is only superficial, partial and fleeting.

The experience of happiness has in common with *enjoyment* the essential feature that its object must always be something good-in-itself. Further, it has for its condition – just as enjoyment does – the "possession," i.e. the immediate, unhindered, unimperilled, intimate contact with a concrete good. It proceeds – as also does enjoyment – with a devotion to the fullness of that good. The experience of happiness is distinguished from enjoyment, as we have already indicated,[21] by the fact that it is not directed to a limited "this, here and now." While, for example, in the case of a gastronomic, aesthetic or intellectual enjoyment, we know approximately what we have to expect from the object enjoyed, and thus already prospectively indicate a limit, in the experience of happiness we push forward, in a determinate direction, into the boundless. In short, enjoyment is an intentional mode of experience, happiness a meta-intentional mode.

The experience of happiness is bound to *joy* by the fact that it is always equivalent to an ever-renewed taking-possession, attaining, discovering. Jubilation, triumph, rapture, emotion can thus be identified in the one who is happy as well as in one who is joyfully moved. But happiness is distinguished from joy by the fact that it does not represent the response of the heart to incipient completion, but to

21. Stephan Strasser, *op. cit.*, pp. 335ff.

completion itself. Indeed, perfect happiness corresponds not only to taking-possession, but also to possession, not only to an endless *adeptio*, but also to an unlimited *possessio*. Thus the moment of subjective certitude, which is so characteristic of the experience of joy, disappears. In its place enters the presence of the good-in-itself, whose subjective correlate is happy devotion.

Happiness, together with *serenity* and in opposition to emotional joy, is characterized as a relatively stable state of the heart. Still, serenity is created out of a "reserve" of satisfying experience. What disposes it serenely is the certitude of the presence of the good-in-itself resonating through everything. Thus it is "consoled." One who is happy needs neither consolation nor certitude, since he experiences the immediate presence of the good-in-itself.

Let us now go on to a positive characterization of the experience of happiness. One of its essential features consists in the fact that its "object" is always unbounded fullness, a superabundance, an infinity of goodness or beauty. It is always something enormous which transcends the measure of the everyday, the usual, the familiar. This is true even of the popular, universally recognized "happiness-producing events." When evaluating it, the phenomenologist, of course, will not inquire what it signifies *sub specie aeternitatis*, but how it is experienced. For example, an appointment appears to promise happiness to me, since it permits me to develop a fruitful, prosperous effectiveness whose possibilities for myself I consider unlimited. Every wedding is, in my eyes, an inexhaustible source of joy and of harmoniously beautiful communion. The son who will be born to me I conceive of as an exceptionally richly gifted being. I expect that he will have success everywhere I have failed: that he will overcome all obstacles which have limited my effectiveness, and so forth. The return from confinement signifies for me, not the end of a miserable state, but above all a colossal resurrection of my existence as an individual and as a member of a community. With the winner in a lottery, I believe I possess unbounded possibilities of gratifying my material desires. The child under the Christmas tree stands before a superabundance of glittering, shining, fascinating, captivating things; he is happy – not because he feels himself rich, but because he feels himself superabundantly rich. What therefore characterizes the so-called "causes of happiness" is the fact that they open up infinite perspectives to us. They allow us to envision a *concrete infinity.*[22]

In the notion of concrete infinity we encounter a serious roadblock. From the purely logical standpoint, it must be rejected immediately as contradictory. The *concretum* is the being which has "grown-together," with its individual, unique, inexchangeable characteristics. Now, it is very easy to show that the category of infinity in no way permits closer determination, much less can it then be the bearer of any individual characteristics. On the other hand, a totality with distinct individual features is quite conceivable; but then this totality could not be considered as the totality-as-such.

22. In the following treatment of the problem of happiness we shall not consider the specifically mystical experience. Since this cannot be considered a universal experience, it is not suited to a phenomenological approach.

It is clear that these arguments from the formal-logical standpoint cannot be refuted. The question is only whether it is advisable to accept this standpoint immediately and to the exclusion of every other approach. Would it not be more correct to attend first exclusively to the phenomena and describe faithfully what presents itself to us? Is it very wise to wish to argue away, at any price, what does not immediately fit into one's pet system? In this way, does one not run the danger of imitating the botanist who, in examining a plant, exclaimed: "There is no such plant!"? What cannot be thought can still be experienced. What the formalistic philosopher discards as contradictory, the phenomenological philosopher may recognize as a piece of evidence – but note well: as a *preliminary* piece of evidence.

Now, it is evident that contradiction is actually contained in the human experience of happiness. The experience of happiness relates, on the one hand, to a totality which always has a determinate nature, an inimitable peculiarity. Happiness is not only the transcending experience of result-as-such, of love-as-such, of superabundance-as-such, and so forth, but that of a completely determinate mode of success, love, superabundance, etc. The warrior would not be content with success on the stage; the actor does not dream of military honor; whoever loves person X cannot "be happy" with Y, etc. Hence the happiness of man is bound to completely concrete forms. On the other hand, happiness transcends all objective limits and categories. For the actor, for example, the applause of enthusiastic crowds means simply everything. For him, the glamor of artistic fame irradiates the firm outlines of objects; for him it is something infinitely meaningful and, at the same time, a determinate infinity. Accordingly, if we speak of the world of one who is happy, then we think, on the one hand, of a world in the ancient sense of the term, i.e. of a *κόσμος*, of a well-ordered, adorned, completed whole. On the other hand, we think of a world in the modern sense, i.e., of an infinite expanse without limits, incomparable in its immensity. None of these two apparently contrary characteristics may be entirely wrong. Only one time the accent can be put on the feature of integral completion, the other time on the immensity of the completion.

A further feature of the authentic experience of happiness which at first seems paradoxical lies in the fact that there dwells within it the claim to "a certain eternity." We are the better aware of the contradictory character of this new conceptual structure because the opposition which is expressed in it also causes tension in experience. In the concrete life of man it is only possible to shift the center of gravity to one or the other of the two conflicting aspects. More precisely, what are these aspects? If, with Boethius and Thomas Aquinas, we attempt to circumscribe eternity conceptually as *interminabilis vitae simul et tota possessio,*[23] the complete essence of *aeternitas* is expressed in the second part of the quote. In the experience of happiness of certain men, the accent falls on this totality. For them, the one who is happy has above all, the character of "full-fillment." But the one who is fulfilled is not the opposite of the short-lived, but of the unfulfilled. That which gives happiness can thus never be an imperfect, deficient half-affair. Universally, one could say that a compromise, a substitute, a merely approximate good or utility never com-

23. Aquinas, *Summa Theologiae*, vol. I, q. 10, art. 1.

pletely satisfies us. On the contrary, what delights us is the finished, the complete, the "per-fect" (i.e., completely realized).

All this is compatible with restricted duration. Consider, for example, an Olympic victor who knows exactly that he will be too old in the next Olympiad to be able still to take part with success in the contests. But this does not change his happy awareness of the fact that this year he is the victor. This claim to fame he has won once and for all. No power in the world can cause what he has brought to fulfillment to disappear. From this standpoint, one can also lay hold of the happiness of Don Juan. For him, only the triumph of conquest counts. For him, the length of the love-relation and of employment does not matter. But the deed of having possessed this woman appears to him as something perfect, unrestorable, definitive. Consider also musicians, singers and actors who appear to take no offense at the fleeting nature of their art. What they strive for is the perfection of the moment. To have once sounded this harmony, to have breathed life into this tragic fiction for an evening, is also something definitive, which can no longer be destroyed.[24] The happiness of such men consists in *the perfect actuation and integral command* (in the *simul et tota possessio*) of a limited temporal whole. That this crystallized piece of eternity melts away in their hands does not distress them. They enjoy the triumph of having possessed it once in a total manner.

Other men experience happiness, above all, as duration. They see in it an "unrestricted life." Correspondingly, we treat, for example, a marriage, the birth of a child, the reawakening of one's country, as a beginning without end, and in this sense as the beginning of a *vita interminabilis*. Both kinds of the experience of happiness reveal actual, only apparently conflicting aspects of eternity.

In yet a third respect, the experience of happiness comes to be treated as a paradox. It is, as we have already indicated,[25] the experience of a joyful taking-possession and at the same time of an enjoying possession. Both contradictory features are, as such, characteristic of the experience of happiness. Happiness consists, on the one hand, of the consciousness of coming-to-be-enriched, coming-to-be-superabounding, coming-to-be-chosen, coming-to-be-elevated which receives plastic form in the myths of Danae and Ganymede; and, on the other hand, in the knowledge of one's own being-rich, being-chosen, being-elevated, etc. Thus, the awareness

24. What has been said here of individual performances, acts, effects, and activities, Rilke employs for human existence as such. It appears to him to be fulfilled on earth:

"....Und wir auch
einmal, Nie Wieder. Aber dieses
einmal gewesen zu sein, wenn auch nur einmal:
irdisch gewesen zu sein scheint nicht wiederrufbar."

Duino Elegies, Ninth Elegy

".... And we, too,
once. Never again. But this
having been once, though only once:
having been on earth appears beyond recall."

(*Duineser Elegien und Sonette an Orpheus* (Wiesbaden: Insel-Verlag, 1946); trans. J.B. Leishman and Stephen Spender: *Duino Elegies* (London: The Hogarth Press, 1942).)

25. Stephan Strasser, *op. cit.*, pp. 342-343.

of happiness combines, at the same time and in the same respect, the experience of completely dynamic becoming and perfectly static being. It displays the features of triumphant *adeptio* and restful *possessio*.

Human existence does not, of course, allow the simultaneous combination of the two conflicting moments; and thus it is that in the concrete human existence sometimes the one, sometimes the other essential feature is more clearly prominent. The restful happiness of the idyllic existence, the satisfying occupation, the harmonious marriage or friendship resembles more an enjoyable possession. Yet the correlative moment is not entirely lacking: profession, marriage and friendship would not be experienced as making us happy if they did not repeatedly afford occasions for discoveries, taking-possession and realization that make us happy. Vice-versa, with the conquerer, adventurer, or performer type, the center of gravity would lie in the *adeptio*. Still, one of these existential types who really is happy is likewise aware that he is able to call his own the victory, the fruitful adventure or the artistic creation, and thus in some way to possess it.

We still do not know how to explain these three paradoxes of the actual experience of happiness. We were only able, up to now, to point to an apparent solution. It would be quite simple if one were able to ascribe the aspect of infinity in the experience of happiness to reason and its concrete aspect to sensibility, and explain that man, because of his spiritual-sensory nature, is able somehow to synthesize these two heterogenous "existential components" into a whole. It is clear that man renounces himself with this assumption from the framework of strata-theory, and exposes himself to that criticism which we ourselves have given to this psychological and anthropological conception.[26]

One finds nothing of this dualism in experience. It is out of the question that the adhesion of the experience of happiness to the concrete is felt as a weakness, an imperfection, a vestige of animality. Rather, we experience the infinite precisely *in* the concrete and *with* the concrete. Precisely that which produces the transcending vision of superabundance is also a concrete thing, here-and-now. For example, the husband who experiences his harmonious life with his wife as a source of happiness would not be in a position, in view of that experience, to look away from her individual uniqueness. Thus it seems to us more honest – and besides, psychologically more correct – to hold that the object of the human longing for happiness exhibits a concrete infinity. Thus, at least the primordial datum upon which all further philosophical questions and probings build, is described genuinely and straightforwardly. The (would be!) explanation of the phenomenon will not be imposed from the outside on the phenomenon itself. By this honesty and frank acknowledgement of being able to see a paradox on a certain level as nothing else but paradox, one guards against superficial apparent solutions and forces thought into the paths of serious searching.

26. *Ibid.*, pp. 149ff. and 243ff.

8

Aspects of the Sexual Incarnation. An Inquiry Concerning the Meaning of the Body in the Sexual Encounter*

JAN LINSCHOTEN

"We are condemned by nature to live in the imaginary,
and in what cannot be completed.
And this is living."

P. Valéry[1]

I

"Among all experienceable things our body is the most our own,"[2] but – amazingly enough – the body has been overlooked by psychology for many years. To be certain it has been spoken of: but merely as *res extensa*, as an apparatus or mechanism, a status it owes to Descartes. However, the body which thus was conceived of and maintained as a "preparation", that is as a thing neither alive nor dead, has been assigned a new importance by anthropology in the past few decades, and has now begun to regain its place in psychology, too, as that with which man lives and in which he lives. The body as *means of eksistence* has become a central theme. In Husserl, however, it is still left out of consideration when taken in this function: phenomenology started with the rediscovery and unfolding of the *Cogito* and therefore continued to consider the body as merely an object of intentional acts: thus the body in its *facticity*. And in Heidegger, too, who shifted phenomenology's center of gravity from transcendental consciousness to being-in-the-world, the body is not mentioned; in *Being and Time* fewer than 10 lines are devoted to the body.[3] Interest in the body, thus, did not originate in phenomenology, but in the philosophy of life, in Maine de Biran and in Bergson. It was precisely the contact between these two currents which made the problematic of the body a fertile field for con-

* "Aspecten van de sexuele incarnatie" appeared in *Persoon en wereld: Bijdragen tot de fenomenologische psychologie.* Utrecht: Bijleveld, 1969, pp. 74-126. Reprinted by permission of the Publishers. Translated for this volume by Joseph J. Kockelmans.

1. P. Valéry, *Analecta* (Paris: Gallimard, 1935), p. 198.
2. F.J.J. Buytendijk, *Algemene theorie der menselijke houding en beweging* (Antwerpen: Standaard Boekhandel, 1948), p. 57.
3. A. de Waelhens, *Préface* to M. Merleau-Ponty, *La structure du comportement* (Paris: Presses Universitaires de France, 1949), p. *vi.*

temporary psychology. On the scene of recent French thought the crop has been harvested by Sartre and Merleau-Ponty. But this does not mean that the chapter on "Bodilihood" is complete. On the contrary, thus far we only have the introduction to the subject.

In the pages to follow I wish to see what the phenomenology of our bodilihood can contribute to offering insight into the meaning of the body in the sexual encounter. To repeat: amazingly enough in the many studies on the psychology of our sexual life we find little – too little – on the body as it is lived there and experienced, although it is almost impossible in this realm to overlook it. We must turn, therefore, to literature, the world of the poets – where the primary Eros as it were comes to self-consciousness[4] –, but especially to the world of the more prosaic authors who have described the sexual encounter and incarnation in numerous aspects. On more than one central point we will have an opportunity to make fruitful use of Buytendijk's reflections on bodiliness for our particular question.

II

Van den Berg to a great extent has adopted Sartre's distinction between the three dimensions of the human body, albeit not without modification.[5] Let us see inhowfar we can take our point of departure in this analysis.

The body taken in the first dimension is a body which is passed over in silence (*passé sous silence*), passed over in the direction of the scene. That is to say, man in action forgets his plans and his body. He is completely taken up by his field of action. It is over there, in the scene, that he finds everything of which he himself is conscious in his actual concern. He is unaware of his body; his "inner constitution" – and the experience of his own body would have to belong to this state, also – may not be sought for *within him*, but out there, outside. The subject shows itself in the things and thus psychology is strictly speaking *cosmology*: description and analysis of the human scene.[6] Therefore we should search for sexuality, too, not in man, but in the *sexual scene*: sexuality "is psychologically not to be found first 'with the subject' nor 'on' his body, but it shows itself as a world."[7] Here we must already note at once the following. There can be no doubt that the world looks different to the two sexes, and that it is in this difference of world that the psychological distinction between the sexes manifests itself and becomes clearly graspable for the first time. But does this mean that sexuality in the narrower sense, that is erotic attraction, sexual drive and sexual desire manifest themselves primarily as a

4. V.E. von Gebsattel, "Ehe und Liebe," in *Zeitschrift für Völkerpsychologie und Soziologie*, 1 (1925), p. 251.
5. J.-P. Sartre, *Being and Nothingness*, trans. Hazel E. Barnes (New York: Philosophical Library, 1956), pp. 306-366. J.H. van den Berg, "Menselijk lichaam, menselijke beweging," in *Nederlandsch Tijdschrift voor Psychologie*, 5(1950), especially, pp. 402-412.
6. van den Berg, *op. cit.*, p. 296.
7. *Ibid.*, p. 405.

sexualized scene? Does this mean, then, that in the sexual experience, too, our own body is passed over?

In order to avoid misunderstanding we must mention that in what follows we are using the expressions "sexuality" and "sexual encounter" in the same sense as Michel has: "The sexual appears for the first time when the subjective drive, which in itself is without orientation, wandering, dark and without form, becomes changed through the encounter with a determinate person of the other sex into the desire to have bodily intercourse precisely with this human being."[8] By this move we have thus already excluded that "sexuality" manifests itself only in the scene. True there are many examples to show that amorousness and desire manifest themselves in the charm of the landscape, the attraction of the woods, in the mysterious, enchanting, lovely, fascinating character of great and small things. We recall the scene described by Gide: as a child he is walking with his nursemaid and asks her suddenly why she is laughing. For nothing, she says, I am just laughing, it is beautiful weather. *And immediately the valley was filled visibly with love and happiness.*[9] But there is no question here of a sexual relationship, only of friendship, perhaps of childish love in which one can and may pass over the body. We have another example in the desire of *Madame Bovary* who, disappointed over her colorless life with her husband, dreams about love and great adventure. It does not amaze us to read that she liked the sea *because of its storms*. Her desire is not yet *sexual* desire. It is not even so when she encounters *Léon* for the first time and the two discover their common feelings in a conversation. When they talk about the sea she says: Do you not think the mind moves more freely about above the limitless space which elevates the soul and inspires thoughts about the infinite and what is ideal? And he: it is the same with mountain-scenery. Such a scene must lead to enthusiasm, prayer, ecstasy.[10] It is in this common, expanded world which frees the young woman and the clerk from their fetters and grants them the freedom they desire, that they find one another, albeit only imaginarily. They affirm their being-together in the delivering landscape; that is why the conversation contains, strictly speaking, a profession of love. Here we understand the meaning of the saying that man passes himself continually by materializing himself in the world of things.[11]

The enamoured desire for the other transforms my world and becomes manifest for the first time in this transformation. Frank Harris describes how this happens in connection with his first, still very youthful love-affair, "now all of a sudden, in a moment, my eyes were unsealed to natural beauties."[12] From this moment he began to enjoy descriptions of nature and landscape-paintings. We encounter here a transformation of nature, at first meaningless, into a *beautiful* nature, in correlation with the awaking but still hardly specified amorous desire. Marcel Proust's

8. E. Michel, *Ehe, eine Anthropologie der Geschlechtsgemeinschaft* (Stuttgart: Klett, 1948), p. 11.
9. A. Gide, *Si le grain ne meurt...* (Paris: Gallimard, 1928), p. 58.
10. G. Flaubert, *Madame Bovary*, in *Oeuvres complètes*, vol. 8 (Paris: Conard, 1910), pp. 112-113.
11. van den Berg, *op. cit.*, p. 295.
12. F. Harris, *My life and loves*, vol. I, (Paris: Obelisk Press, 1949), pp. 57-58.

observation is already much more advanced.[13] "But wandering through the woods of Roussainville without a country girl to embrace, that was tantamount to not knowing the hidden treasure and profound beauty of these woods." "... the passing girl who evoked my desire appeared to me to be not any one exemplar of that general type: woman, but a necessary and natural product of the soil."

Desire for a more intimate contact with a member of the opposite sex becomes manifest to Marcel in the character of the woods: mysterious and fascinating woods with hidden treasures. And these woods are such an erotic landscape that, as it were, it automatically evokes the woman as a natural product of the soil. Conversely, it is only now that a real encounter with a real woman, which corresponds to the desire covered in these woods, can reveal the hidden secret.

The melancholy recollections of Stendhal come alive in the landscape without his being involved in it in his sexual bodilihood: "The line of the rocks which, I believe, border on Arbois and via the main road come from Dôle were for me a sensible and evident image of Métilde's soul."[14]

There is no doubt that these "innocent" love-desires manifest themselves in the landscape while the body remains "out of the picture." But it is precisely the fact that the body is scarcely addressed in its *sexuality* that defines them as "innocent" encounters. Where love-desire becomes *sexual* desire, one's own body immediately comes into play and enters the experience. We see this in the following passage from Camus:[15]

> There is only one love in this world. To embrace the body of a woman is also to hold close to oneself that strange joy which descends from heaven toward the sea. In a little while, when I throw myself down on the absinthes to make their perfumes enter my body, I shall be aware of accomplishing a truth which was that of the sun and which will be that of my death, also.

It is here that for the first time a sexual desire addresses itself to us and this cannot be except by incarnating itself in the body.

The booklet from which this passage was taken is called *Wedding* and it bears this title with good reason although it does not describe the wedding of a man and woman, but the erotic unification of man and the landscape. Sexual desire can originate when a real partner is absent, it can manifest itself in the strange joy which descends from heaven to the sea; but this cannot happen so long as one does not want to throw himself onto the flowers in order to let their perfumes penetrate the body. Sexual desire manifests itself in the landscape, but this landscape immediately motivates an incarnation of my sexuality. One can only experience this desire in a dialogue between landscape and body. Elsewhere we read:[16] "On emerging from the commotion of the perfumes and the sun, and into the air which is now

13. M. Proust, *Du côté de chez Swann*, vol. I (Paris: Grasset, 1919), p. 146.
14. Stendhal, quoted by S. de Beauvoir, *Le deuxième sexe*, vol. I (Paris: Gallimard, 1949), p. 365.
15. A. Camus, *Noces* (Paris: Charlot, 1947), p. 18.
16. *Ibid.*, p. 23.

cool because of the evening, the mind becomes calm, the relaxed body tastes the inner silence which flows forth from the satisfied love."

Here it is still barely possible to distinguish between the bodily event and the happening in the landscape. The inner silence, the relaxation of the body forms one pole of a happening that can be equally well described as the silence of the nightly landscape and the relaxation of the night.[17] Sexual desire originates in an *encounter* in which the body is not passed over because it plays an essential part in the encounter. This is also indicated by Sartre – following the description of the three dimensions – when he says that I am an accessory to my desire; or better, the desire has fallen wholly into complicity with the body.[18] The sexual desire *is* an attraction between the sexual landscape and the person who is bodily related to it. Properly speaking one cannot say where the desire emerges perceptibly except in this desiring dialogue itself to which I am an accessory through my body. Julien Green with a master's hand describes the awakening, still very vague sexual desire of a young girl in her puberty:[19]

> The heat of the bed woke me up. The comforter weighed so heavily on me that I believed there was a body lying across my body. I rose, half-opened my nightgown which the sweat had glued to my throat and headed, feeling my way, for the window. A warm breeze blew in from the garden; in order to breathe it better I pushed against the heavy wooden shutters. The air lifted my hair on my forehead; with my arms on the edge of the window-sill, I looked at our plane-tree asleep in the moonlight and at the lawn rimmed by the strange light which fell on it. *This spectacle intoxicated me.* It seemed to me that I was going to discover *the secret aspect of things*, those which one catches a glimpse of in a dream, but which one forgets when one awakens, and at the same time *I felt myself being overcome by an immense tenderness*; yes, I lowered my eyelids and offered my mouth to the night, to the wind which ran over my face. *A desire came over me to sway and to hold out my arms.* I did not recognize myself. I was going to find myself confronted with something new. The world as I had seen it until now lost its power, its reality: in its place the night held sway. I had never been permitted to look at the night, to love it. From a certain hour on my life had to annihilate itself in sleep. Who could say what was going on around me now? From pure pleasure, I started to laugh aloud and you must believe me, jumped up and down while holding onto the edge of the window-sill. I was free, or at least I forgot my fears, concerns, and all my ties with the preceding day.
>
> The exaltation of this minute inspired me to a peculiar gesture about which I myself was surprised: I took off my night-gown and looked at my body.... I admired the whiteness of this flesh which I had never seen; my

17. Cf. my article "Over het inslapen," in *Tijdschrift voor Philosophie*, 14(1952), no. 2 (Cf. this volume, pp. 79-117).
18. J.-P. Sartre, *op. cit.*, p. 388.
19. J. Green, *Le visionnaire* (Paris: Plon, 1950), pp. 50-51. (Our italics.)

> hand which placed itself on it received a sensation of delightful freshness, and I asked myself why this would be bad.
>
> If I remember it correctly I then stretched out on the floor, my belly and upper legs suffused with an accessory light. Normally I was unable to stand the darkness of a room without a light, but this night I became different. I laughed aloud from the sheer sense of well-being I felt while rolling on the carpet like a young animal feeling its strength.

There is a simultaneous conversion of landscape and body. The girl discovers something new: a garden in the moonlight, an old tree standing in that capricious light; she does things which amaze her: she stretches her arms, kisses the night, looks at and touches her body. She bares that body so that light as an accomplice shines on it, and she changes. A desire which addresses itself from the landscape, too, becomes incarnated in her body. Just as the night manifests its hidden aspects, so does her own body appear to her in a new meaning. The conversation with the night motivates the dis-covery of the body just as does the pleasurable rolling on the carpet during which she first really experiences the scattering desire.

When there is no question of a passing over, but rather of a dialogue, we are already confronted with the second dimension of the body. That is to say with the body as it appears in the eyes of the other. Someone else; in Green's passage it is the girl *herself*: she begins a dialogue with her own body. It is *in the absence of someone else that auto-eroticism originates*. But before going into this we must first concern ourselves with a general consideration of the second dimension.

The mode of appearance of the body in the eyes of a spectator is the second ontological dimension of the body. Whereas the busy person forgets his body and passes over it in the direction of the landscape by which he is taken up and with which he is concerned, for the spectator there appears a human body *which moves*. He witnesses how from the body as the active center of the landscape all kinds of intentions flow forth in the form of acts which find their end points in the landscape. The other appears as meaning-giving body insofar as I understand his acts: he drinks a glass of water, or he reaches out for it, etc. But this implies that the other appears as *situated body*[20] and that it is this situation which makes his activities meaning-giving for me. In order to understand what someone does, we focus our attention on his situation in order to understand him from it. Thus I see how he himself passes over himself. *But simultaneously I see that "he himself" escapes me.* Insofar as he hammers, saws, or writes, etc., he is clear to me; but as soon as I look at *him*, as soon as I realize that these activities flow forth from *him*, and so are guided by an 'inner center", I experience a *hidden interiority.*[21] There is obviously more there than just that body, there is a "personal psychic life" in which I do not take part. The precise advantage in the situation where I, as an unnoticed observer, can observe the other is that I can ask myself quietly – or uneasily – what *really* is going on in him. I can grasp his inner life only insofar as it flows through his body and

20. J.P. Sartre, *op. cit.*, pp. 344-345.
21. Cf. van den Berg, *Menselijk lichaam*, pp. 410, 423.

manifests itself in his dialogue with the landscape. I discover his plans insofar as he executes them.

We may now leave this situation and consider the whole: a busy man, and an unnoticed observer who spies on him. We see that here, too, there is a question of an *encounter*. The person who is spied on is taken up in his landscape. The spy is equally so; the only difference is that the latter's scene contains still another man. *But the passing over of one's own body and one's own interiority is here never complete.*

The deficiency of Sartre's analysis is (1) that he considers the extreme situation as a genuine possibility and thus thinks that the description of the person in the first dimension *coincides* with the description of the landscape; and (2) that he *isolates* the body in the second dimension as the body of the other. That is to say he does not sufficiently take into account *that I am spying on the other*. It is not just a situated body, but a body which is situated *for me*, the secret observer who is involved in the situation as a whole. *My being-a-spectator has been incorporated into the mode of appearance of the other*; and this is true even if the other did not notice me. It is precisely the latter which motivates my peace of mind while I am spying on him; he does not see me. But this is a peace of mind which implies that I myself, while spying, do *not* completely pass over myself. Think of the "pounding heart" which we most certainly experience in *ourselves* when there is a chance that we shall be caught spying. Sartre himself has certainly seen this since in passing he says: "I never apprehend the other as body without at the same time in a non-explicit manner apprehending my body as the center of reference indicated by the Other."[22]

My being-an-observer is codeterminative for the manner in which the other appears to me. When Susanna takes a bath this activity need not imply any erotic constituent for herself. But when I spy on her in so-doing her body sexualizes itself in my eyes. The manner in which she moves her own body, perhaps just one single gesture, makes me immediately and spontaneously suppose an interiority in which all kinds of things are going on. For whom does she make herself so pretty, etc.? Furthermore the body of a beautiful woman seems to bear witness to a mysterious interiority; we shall return to this subject later. But I am not a disinterested observer here, completely passing over myself in the direction of the erotic landscape. I constitute it as an erotic landscape *through my own spying* in order then (that is simultaneously) with this landscape to *motivate my sexual incarnation*. Henri Barbusse has illustrated this happening in detail in the story of a man who while in his hotel room spies on his neighbors through a hole in the wall. Everyone who knows this book knows how just by reading it – this literary spying – an almost unbearable sexual tension can arise as he lives the story. When our hero spies on a woman who undresses in the semi-darkness, we read: "And my flesh cried branded as if with a red-hot iron by the voluptuous line which, growing, disappeared into the darkness, vanished into extraordinary depths."[23] In the undressing as such there is

22. J.P. Sartre, *op. cit.*, p. 344.
23. H. Barbusse, *L'enfer* (Paris: Michel, 1949), p. 34. English translation by R. Baldick (London: Chapman and Hall, 1966), p. 25. (Our italics.)

nothing sexual; it is the desire-filled spying which immediately constitutes the sexual character of the landscape as well as the sexualization of one's own body.

The manner in which the voyeur turns toward the other, in a onesided attitude which commits him to nothing – for the other does not see him – becomes directly expressed in one's own body. For as we have said already, if the other person is lacking as *partner*, then one's own body becomes a sexual dictator who asks *to be dealt with*. The other person is, indeed, present here objectively, but not as partner. My sexual relationship then becomes a relationship with my own body instead of a relationship with the other. This presupposes a certain *alienation*. That is why we often find the testimony that one's own body can fascinate as if it were a stranger. In this turning back toward one's own body the body receives a certain "independence": this is indeed I, but at the same time it is not, it is someone else, a living and intriguing body with its own hidden interiority which is not mine. It is particularly during puberty, and especially in the girl, that we find this "narcissistic" attitude as a natural attitude. As for puberty, we saw an illustration of this in Julien Green. The love-desire arises as desire for a partner. At the same time the young person experiences his own body as that through which the relationship with the other must come about. But he, or she, is still *immature*: does not see an opportunity for realizing this "mature" relationship with someone else. In the meantime he *asks* already, his body becomes transformed and prepares itself. Perhaps he finds a way out by provisionally accepting this fascinating body of his as a "partner," to admire it, to cherish it and to take care of it, to experience its mysterious attraction. Masturbation is nothing surprising in this connection.

This is completely different in the mature woman. Already through her "caring" attitude which is characteristic for her *as* a woman, she is inclined to include her own body in her caring. But in addition to this – we shall see later why it is so – the female body is that which is pre-eminently sexually fascinating, also to the woman herself. In the novel *Aphrodite* there is a long passage in which this ecstasy in regard to a woman's own body is described: "With her hair and her breasts she made a thousand charming games. Sometimes she even accorded a more direct satisfaction to her perpetual desires."[24] And it is precisely the bathtub in which she sees herself, that the courtesan considers a suitable place for this calculated satisfaction. Masturbation, too, is not an activity-in-isolation, but a dialogue which already begins with the narcissistic care of one's own body. But in a mature woman it certainly does not have to come to that. Often we find no more than a secret – or outright – pleasure in her own body. There is nothing strange in the fact that Madame Bovary is extremely aware of the delight she experiences in combing her hair which falls down far over her shoulders.

But even if the partner in the sexual encounter is not a *sharer*, the encounter is still a dialogue, a dialectical relationship and not a keeping oneself aloof from, and passing over of, oneself on the part of the experiencing subject. The complete passing over of one's own body is a theoretical limit which could be realized if a

24. P. Louys, *Aphrodite* (Bruxelles, 1946), p. 22. English: *The Collected Works of Pierre Louys* (New York: Libra, 1932), p. 17.

man could behave purely pre-reflexively.[25] In the concreteness of our experiences we always find a talking-back and a reflection, an experience of my own body, albeit it perhaps merely a secret knowledge of myself insofar as I am bodily involved in the situation. That is why we are unable to understand the sexual happening through an analysis of the landscape alone, but rather through an analysis of the sexual *encounter*. However, encounter is impossible without the body, and equally not without a "notion" of some kind concerning my own body as the medium through which the other encounters *me*.

III

My body expresses the manner in which I encounter the other; but this does not yet mean that this expression *determines univocally* the way in which the other understands it. What my behavior, that is my posture, gestures, and expressive movements, mean *for the other* depends also on *his* "attitude," that is on the manner in which he enters the situation and encounters me. If we are to analyze the sexual encounter, we must first occupy ourselves briefly with the encounter in general and subsequently with the ambiguity and unambiguousness of the behavioral pattern.

We become present to one another as persons *through our bodies*. Or, as Buytendijk has expressed it: "It is through the body that a man is present in the body."[26] That is to say, when we encounter a man, we do not encounter him as someone who is enclosed in a body, is present *in* it, but through his behavior, through the postures and movements of his body, through his expression the other is present to us through his words without ever *being* present completely, as a thing. Buytendijk's reflections are inspired by Heidegger's explanations of the becoming-present of a god in the temple.[27] The god is in the temple into which we enter in order to encounter or adore him, not present as is a thing which *is* or *is* not there, but his "presence" is properly speaking *a becoming-present*: "*Anwesen*". This being-present which really is a becoming-present is thus a happening, but never a completely fulfilled presence. There is no question of a presence which at first is only partial and then becomes more and more complete. It is precisely no more than a *happening which is.*[28] The god becomes present to us *in* the temple, *through* the temple which mediates his becoming-present.

In a similar manner Buytendijk explains the encounter between man and man; however there is an important difference here. First of all, the temple is a building-structure, *something made* with the intention of letting the god "become-present"

25. Cf. M. Maurice Merleau-Ponty, *Phenomenology of Perception* (New York: Humanities Press, 1962).
26. F.J.J. Buytendijk, "Zur Phänomenologie der Begegnung," in *Eranos-Jahrbuch*, 19(1951), p. 467.
27. M. Heidegger, "Vom Ursprung des Kunstwerkes," in *Holzwege* (Frankfurt a.M.: Klostermann, 1950), pp. 31ff.
28. Buytendijk, *Ibid.*, p. 465.

in it. I have to understand the meaning of the temple in order to understand it as a *temple*; in this fact the condition for encountering the god through the temple is fulfilled. On the other hand, the body is not something made but is from the very beginning a *human* body, a dynamic medium through which the *person* can manifest himself. Whereas the temple is made once and for all and thus completed, the body is continuously "made," that is animated. Secondly, we do not encounter the temple in that the latter does not encounter us; through the temple we encounter the god if we *believe* that he becomes present there and in turn encounters *us*. The encounter thus presupposes two persons. When these two *mutually* "become present" a fulfilled and *authentic* encounter realizes itself, one in which *both* of us *take part* in the situation and thus are together. Let us recall the secret spectator for a moment, now we can understand that he encounters his "partner" inauthentically: only the other becomes present to him, but not *vice versa*; the mutuality is lacking.

According to Buytendijk a phenomenology of the encounter must explicitly trace the bodily particularities which must be given so that the person becomes present in a determinate way. *A phenomenology of the encounter is thus at the same time a phenomenology of behavioral patterns*. In different situations man has an "other" body, and thus "is" an "other": he becomes present in different ways depending upon the character of the encounter, *and this character of the encounter becomes manifest in the manner in which the body expresses it*.

Thus it becomes clear to us that in an encounter it never happens that an existing psyche *meets* or *discovers* another existing psyche via the expressive movements of their distinct bodies. The psychic taken as "inner thing," as "not spatial subject" does not *exist* at all. It is only in the encounter that the other becomes present to me through his body, just as I become present to him. We must understand this correctly. Whereas Sartre puts the stress on the body in the encounter, we understand the encounter here as a reciprocal *incarnation*. The body is not only meaning-giving body through the situation, but it is also the medium through which the other becomes present to me. The other *is* not this body, but he incarnates himself through this body. *And that is why each encounter is ambiguous*. The person can never completely *coincide* with his body, *although he exists only through it*. It is an essential characteristic of human bodilihood that the person indeed *is* this body, but at the same time also *has* it, that he is it by having it and has it by being it: his bodily being is being a body by having it. That is why as a person his attitude toward his own body is always ambiguous. His becoming-present is an incarnation for himself and for the other; but the incarnation is never completed, he is never present in the body as a thing (*dinghaft*), he is continuously occupied with incarnating himself. Precisely the fact that I encounter the other as a *person* through his body, means at the same time that a "part" of his personality escapes me. This escaping "part" constitutes itself in my view as his *inner life*. And how easily do we consolidate this hidden interior into something which is in itself closed and limited.

With these reflections Sartre's analysis of the first and second dimensions receives its real meaning for the first time. In the first dimension the awareness and experience of one's own body is not explicit, and yet pre-reflectively it is experienced as the center of my scene, as the source of all activities, as that with which I am in the situation. In the second dimension the body of the other is experienced as the medium through which he becomes present, and not primarily as living and moving *body*. Properly speaking Sartre's analysis of the second dimension is thus founded on an inauthentic encounter: on a situation in which the other becomes present to me, but I not to him.

Let us now consider the third dimension of the body in the light of what has just been said: the body *under* the eyes of the other.[29] What is meant here can be illustrated as follows. While someone is occupied with something he suddenly discovers me as a spectator – according to Sartre principally as an unkind observer. His reaction to my hostile look is that he, as it were, stands there paralyzed. The scene with which he was completely taken up a minute ago, with which he was in dialogue, which belonged to him and surrounded him, making him the center, is now stolen from him. All meanings flow away from him toward me, toward me who constitutes the scene as *mine* through my look. The other is hurt, my look hits him in his body, and he experiences his body as an object-for-me. And so he reacts with confusion, with embarrassment in the broad sense of no longer knowing what to do with his own body. He is standing there without a scene to guide his movements; he no longer knows how to move and his movements become awkward and stiff.

Van den Berg correctly observes that the look of the other does not *necessarily* steal my world away from me and alienate me from my body. For we also know of the *stimulating* look of the kind observer whom we for instance encounter among the rooters at a sports field.[30] Sartre overlooks this, for him the relationship with the other is basically hostile since each of us exists merely for himself. There can be no question of a friendly conversation; and what is more even the nature of the conversation is alien to him. We could perhaps formulate this as follows: in Sartre the encounter is a combination of twice the second dimension; I can transcend the transcendence of the other or choke it down;[31] for him the same thing holds in regard to me; thus the "encounter" consists in a mutual attempt to overpower the other.

The stimulating look enables me to have a better dialogue with the landscape, makes my movements smoother and more appropriate. For the observer who enters the intimacy of the encounter between landscape and me is not a stranger who breaks in, but someone who once again affirms and consolidates this encounter. This can be the case also for Susanna in the bathtub who suddenly experiences the admiring look of her lover. Her reaction expresses itself immediatley in posture and movement of her body. He finds *his* admiration returned in *her* movements. It is possible, also, for her to withdraw from his admiration and eventual desire, or just

29. Cf. Sartre, *Being and Nothingness*, pp. 351-359; particularly the passage on "the look," *Ibid.*, pp. 252-302. See for a very brief summary of these ideas: Merleau-Ponty, *Phenomenology of Perception*, pp. 166-168.
30. van den Berg, *Menselijk lichaam*, p. 412.
31. Sartre, *Being and Nothingness*, p. 363.

simply to pass over these. In the first case the observer reads her indignation in her posture and expression; in the second case he experiences perhaps what Flaiano described: a young officer approaches a girl who is taking a bath in a river and inquires about the road; silently he really asks her something else, also. She shows him the road and equally silently answers his secret question through the manner in which she behaves: "I did not see her naked body but felt her indifferent *breast* close to my shoulder."[32]

When the other's look steals my world away from me, and when he hands my scene to me once again and in so doing dedicates my body to me, in each case there is a question of an authentic encounter, of a fulfilled reciprocity; and this at any rate distinguishes Sartre's third dimension from the second.

And yet the encounter often remains ambiguous – and even has many meanings. For the manner in which the other becomes present *for me* is not univocally determined by *his* "attitude," but begins to exist only in *my* look. I can make a mistake, I can read an inner disposition into his gestures which does not agree with the manner in which he experiences his involvement in the situation. We can now make a distinction here between ambiguous – strictly speaking having many meanings – and unambiguous behavioral patterns. This holds true for the activity as well as for the expressive movement. In the first case – we discussed this earlier – the intentions can be different than I read them from the activity. The meaning which the activity as a perceptible behavioral pattern has for me, and the intention from which the activity flows forth, need not be in perfect agreement with one another. The activity of the other is always at least partly unclear.

We are particularly interested in the second case, in the expressive movements. For these, too, it holds good that the perceptible meaning need not correspond to that which "is expressed." But although it is true that the observer can have his doubts about the *specific* meaning of a determinate expression, the *basic meaning* is clear to him without further thought. Each expression is simultaneously unambiguous – namely in its meaning-nucleus – *and* ambiguous, namely in its concrete meaning for the other. We can illustrate this using the example of *blushing*.

A girl stands combing her hair in front of a mirror and suddenly discovers a spectator whom she prefers not to involve in this activity. What kinds of reactions may we expect here? In principle an infinite number of them. However, we shall limit ourselves to just one. The girl blushes and in this way gives away the fact that she is affected by the immodest look. The observer understands from this blush that she has been hurt as a person, is embarrassed, experiences herself as being looked-at. If she speaks, she may perhaps stutter. The blush is the typical reaction to the *Sartrian* look. The one who blushes, blushes with the body he is, which is a quasi-object for the look of the other (in contradistinction to Sartre we are convinced that the human body can never completely become an object for us). Darwin assumed that the blush emerges through the fact that the attention which the one who blushes focusses on one of the parts of his own body diminishes the tone of the capillary vessels; these vessels are then filled up with more blood than would normally be the

32. E. Flaiano, *The Short Cut* (New York: Pellegrini and Cudahy, 1950), p. 19. (Our italics.)

case.[33] However it is not *one's own*, but the *stranger's* attention which becomes present through the eyes of the spectator, thereby motivating the blushing: "it is never the *eyes* which look at us; it is the other as a subject."[34] One blushes in the face because one is usually looked-at in the face; the face is the habitual striking plane of the encounter. Blushing is a *response*; it would lose its meaning if it appeared in parts of the body which are invisible to the spectator. If parts of the body other than the head are bared, then the blush can occur there also, as Darwin already observed. A pianist blushes in her hands, female patients on their breasts and abdomens, a girl who after resisting for a long time finally became a model even blushed over her entire body.[35] Blushing is something one does, when (correctly or incorrectly, be it morally or bodily) one is suspected of something. One becomes present to the other blushing as the one who is looked-at and *is deprived of the free disposal of his own body*. This is therefore the meaning-nucleus of the expression. This basic meaning is immediately evident to the one who blushes as well as to the spectator. But the *concrete situation* and its future development are by no means determined by it. For the blush taken in itself as an expression does not have a *precise* meaning as the expression of the "state of mind" of the other, and how the observer will interpret it depends in part on the further development of the encounter, the further development of the behavioral patterns, and in part on the manner in which through these forms of behavior the "inner constitution" of the other develops before the eyes of the *spectator*. Is it a blush of shame, diffidence, or embarrassment? Does it mean the beginning of anger or is it already a capitulation? What is going on at this moment in the one who blushes? How easily can one be mistaken in this! The emerging blush does *not yet* mean *more* than being-looked-at, and becomes more precise only in the further development of the encounter. This is what we mean by the "ambiguity" of the expressive movements. There is no doubt about the meaning of the "indifferent breast": the girl rejects the desire of the officer. But it is not certain that she really means it; it can also be an acted or pretended rejection.

In the blush a situation constitutes itself in which those who encounter one another become present to one another in a manner which is evident in its meaning-nucleus. I have just encountered Miss X who (in the course of her personal history here and now) makes herself be present as blushing. She herself as a *person* is not *present* but becomes present *through* her posture and gestures; through the blush as the one who is looked-at. Thus it is in the essence of the encounter that the meaning of the body is enclosed as the medium through which the person at the same time reveals and conceals himself. We must therefore consider the sexual encounter in this light.

33. Ch. Darwin, *Het uitdrukken der gemoedsaandoeningen*, trans. H. Hartogh Heys; *Darwins biologische meesterwerken* (Arnhem-Nijmegen: Slaterus, n.d.), p. 350.
34. Sartre, *Being and Nothingness*, pp. 276-277.
35. Darwin, *op. cit.*, pp. 330, 329.

IV

In so doing we can again take our point of departure in the blush. Blushing is certainly not a specifically erotic reaction. It can occur in thousands of a-sexual situations in which it can become specified as a blush of anger, a blush of shame, etc., through further expressions, gestures, words, and activities. What specifications do we find in the erotic blush?

Because the woman feels herself being "grasped by" the body, she blushes over the indiscreet or desirous look which via her body already has power over her *self*, through the "ambiguous" insinuation or unwanted gesture of approach which she nonetheless cannot resist. The erotic blush is not a *consent* to the desire which speaks from the look of the other, but it is equally not a *rejection* of it. And if the other now in turn begins to blush and gives it away that he feels himself looked-at through the fact that she feels herself looked-at by him and perhaps even suspects on his part all kinds of secret intentions, – then a mutual embarrassment emerges. If an encounter develops from this, then the mutual blush develops as an *ambiguous relation. It is this ambiguity which first discloses the sexual relationship in the narrower sense.*

Ambiguity means more here than just the "ambiguity" mentioned earlier. How can we make it more explicit?

Already in the non-sexually colored encounter between the two sexes we almost always find a striking ambivalence in the reciprocal behavior, observes Buytendijk who explains this phenomenon in part from the difference in self-conception and world-project of the two sexes.[36] But this does not suffice for our purposes. We wish to penetrate more deeply into the phenomenon. In anticipation we can establish already that the part of the body involved in the sexual encounter is ambiguous: *it functions at the same time as medium for the dual incarnation as well as a barrier which makes a complete unification impossible.*

The blush like the smile or any other expressive gesture has many meanings if taken in its concrete signification. The erotic blush, and generally the erotic expression, *remain* ambiguous. This means that the erotic expression in the concreteness of the encounter is characterized by its *ambiguity*: it refuses to specify itself further because in this situation the other as well as I suspend the continuing explanation. There is a reciprocal embarrassment and with this a mutual lack of decision to further specify the situation through a univocal position-taking.

In other words the sexual encounter sets in when reciprocally the expressions of the other function merely in their meaning-nuclei. At the same time there is the mutual desire "to get further." The erotic situation invites one to make a decision without making the decision easily possible. The erotic expression thus becomes specified as erotic through its *unspecificity*, or better through the ambiguity of the encounter in which the expression functions merely in its basic meaning. Why is this so? In order to grasp it we must take our point of departure in the expression

36. Buytendijk, *Begegnung*, pp. 480-481.

and from there ask for the ground which makes the sexual encounter an ambiguous encounter.

Let us consider the very beginning of being in love. This is a situation in which we find the ambiguity materialized pre-eminently. Is it not the typical characteristic of this situation that it is alarming and charming at the same time? Two people sit there together, involved in a wonderful conversation of which an outsider cannot make head or tail. They lose the thread of their conversation and do not finish their sentences. The one says something, the other catches it and tries to understand it in a certain way, but does not feel sure of himself. Has he indeed understood the correct meaning? In each word he searches for a suggestion which will assure him that he can continue to speak in the direction of the decisive word which, on the other hand, he would like to suspend. The mutually flourishing smile marked by this situation increases the uncertainty. What does this smile mean? They do not know. It is perhaps better to postpone the decisive word until the next encounter. What is specific for this situation is to be found in the non-specificity of the expression. The situation is a field of open possibilities. It is useless to ask "what the other means," "what is going on in him or in her." There is nothing but uncertainty which manifests itself in the "openness" of the field of the encounter.

Perhaps we can explain Don Juan's success from the fact that he immediately uses this situation in his benefit, from the first moment in which he encounters the woman eye to eye and the open field of the "Eros of immediacy" emerges.[37] For, when the situation constitutes itself in this ambiguity of postures and gestures, it means already this "immediate Eros," and it is possible through a subtle provocation to confirm it in a beginning love-relationship. Love at first sight is less rare than one thinks. It comes about through a slight modification which suddenly fulfills the sympathetic emptying of the field of common forms of behavior with the decisive word.

When this word is spoken, it means the common decision "that we wish to belong together"; but at the same time it confirms the ambiguity of the love affair and eventually of the sexual encounter. Each time the lovers meet one another their encounter remains under the sign of ambiguity although they have decided together to love one another. This is true not only for amorousness but also for any sexual relation – if we may leave out of consideration here the "abnormal" and "perverse" relationships.

Why is this so? In the human love-relationship and equally so in the less "romantic," "purely" sexual relationship a *game-element* is essentially implied. It is from this game-character that the love-play in the narrower sense derives its meaning. People have tried in vain to compare this love-play – the so-called "preparatory activities for sexual intercourse" – with certain instinctive forms of behavior in the animal. We refer here to what the Germans call "*Balz*" (mating), a prelude which prepares and leads to copulation and according to Bierens de Haan can be taken as an expression of sexual excitation.[38] However these phenomena are merely – albeit

37. von Gebsattel, *Ehe und Liebe*, p. 250.
38. Cf. J.A. Bierens de Haan, *Die tierische Instinkte und ihr Umbau durch Erfahrung* (Leyden: Brill, 1940), pp. 205-212.

impressive – anticipations of the pairing, and strictly speaking, form its beginnings. They find their meaning in the pairing to which all the activities of the partners are related from the very outset. In other words it is not a *free*, but a determined "game" – and therefore much less a game than that of a young animal. In the animal the love-play is really *pre*-play.

In man, on the contrary, the so-called prelude is indeed love-*play*. It has an autonomous function. The erotic activities have a meaning in themselves which cannot be derived from intercourse as a biological factum. If one observes here that the *Balz* has as little a biological meaning as the prelude in man, then the answer must be that this merely *seems* to be so, namely if one separates the prelude from the copulation proper. Yet the *Balz* is an *integral part* of the copulation. But this by no means can be said of the kiss, caress, and of petting. However, we must dissociate ourselves here from the biological viewpoint and ask for the meaning love-play has in the encounter of the "players."

Intercourse *can* be performed without the love-play in the narrower sense. But this implies, at least in certain forms of culture, an offense to the woman. For in a prosaic just-merely-coming-together the sexual encounter is devalued into a "purely biological event." But this is only seemingly so. For even the most impoverished sexual relation, and even the so-called "animal" sexual drive of the one who is sexually "starved" is oriented toward another human being as a partner. Nowhere is this seen better than in the so-called perversions such as sadism, masochism, fetichism, etc. The *mode* of participation may be varying, but the essential relatedness toward the other remains the constant element. It is this relatedness toward the other which founds the playing character of man's sexuality. Where it is a question of the other, or better of a relation from me to the other which materializes in an encounter, the sexual relationship becomes elevated above the biological level. Not in the sense that something becomes added to the biological happening. There is an original constitution of meaning involved which should not be understood as an addition sum made up of biological factors and human fantasy.

However, we must ask ourselves what is meant here by "play-character" and how this conception is founded in the nature of the sexual encounter. According to Buytendijk it is the Gestalt which appears and unfolds itself in encounter and intercourse, which, in its *image-character*, invites us to play. This characteristic of the play-*thing*, the toy, receives a completely new and richer meaning when we apply it to love-play. Let it be stated first that along with Buytendijk we admit an erotic primordial form of playing (in addition to an aggressive form): a circumspect grasping oriented toward a yielding experience of the quality and form of the other.[39] The body of the other becomes here in the playing-petting a play-thing: breast, lips, and skin in their materiality. It is important to understand this correctly. In erotic play the body, as it were, is taken "ironically," *as if* the other does not incarnate himself therein; thus one can play with the fingers or the hand of the beloved as if they were play-things. We now place the play-character of the sexual encounter opposite this erotic form of play. It is obvious that in this context

39. Buytendijk, *Begegnung*, pp. 449ff.

"play" no longer means "a pleasant or exciting occupation for the purpose of amusement or relaxation," or "the opposite of seriousness." "Play-character" means here: *between* play and seriousness; it represents the character of so many human forms of behavior which must be evaluated as *superfluous* the moment we wish to consider them biologically.[40]

Love-play is not enacted *in order to play*; it is play not insofar as it is taken as lived experience; rather it is play in its essential structure as encounter.

In sexual "play" the body is not taken as "thing-image," "toy," in the as-if-form, but rather as the medium through which the other really incarnates himself in amorous play. Love-play is first and foremost oriented toward the other, and not primarily toward the *body*. Erotic play aims at the body as image in the sense of quasi-*thing*; in sexual "play" this body functions as an image of the person, of that "inner form" which incarnates itself in the perceptible form and can manifest itself only via this image. Desire reaches for the person in his totality, and not merely for the person who becomes present here and now. One single quotation from Proust's works may illustrate this. Love does not emerge and cannot maintain itself except in case "one part remains still to be conquered. One loves only what one does not yet possess completely." Elsewhere he writes: "I had not wished to undress her to see her body, but in order through her body to see all the notebooks full of recollections." Or also: the image which I was looking for was not Albertine with a life unknown to me, but an Albertine "who would be known to me as much as possible";[41] "her whole life ... inspired me with desire; a painful desire, because I felt that it was unrealizable, although it was intoxicating...."[42] These quotations could be amplified with countless others. They rest on an experience known to everyone. In the somewhat more pessimistic language of Sartre – a language we must be able to look through while reading – it is said: "The *sexual desire* is my first original attempt to get hold of the other's free subjectivity through his objectivity-for-me (that is his body)," an attempt which in his view essentially and necessarily ends up in "dupery."[43] We must note here that Sartre speaks of "the sexual desire" and in so doing confirms that what has been said thus far is true for more than amorousness or "romantic love." The incarnation of the other (and of myself) stands at the center, even at the center of the most "loveless" sexual desire. Sexual desire "intends" total incarnation.

But if the issue here is about the other in his totality, then it is clear that he or she can never become present completely; for he has his own history which is never completed, his hidden inner life, that secret hinterland which I can never enter. That is why even in the most intimate relationship something belonging to the other keeps escaping me; *and perhaps this is what is most essential*: that which

40. Cf. my essay "Over de humor," in *Tijdschrift voor Philosophie*, 13(1951); also: F.J.J. Buytendijk, *Het spel van mens en dier* (Amsterdam, 1932).
41. M. Proust, *La prisonnière*, vol. I, (Paris: Éditions de la Nouvelle Revue Française, 1923), pp. 91, 115, 131.
42. M. Proust, *À l'ombre des jeunes filles en fleurs*, vol. I, (Paris: Éditions de la Nouvelle Revue Française, 1920), p. 88.
43. Sartre, *Being and Nothingness*, pp. 382, 376.

makes him or her this unique person, this "personal secret"[44] cannot and should not be made explicit. Only a desire for intimacy which is misunderstood searches for a complete *baring* of the other. The intimacy of the sexual encounter respects the personal secret; in this being-together the revelation of the other is merely intended as *desired* revelation. In the sexual encounter the person does not wish to eradicate his desire, but to realize it *as* desire. *That is why* we call this encounter play. It is pre-eminently a song of desire. That is also why the becoming-present in the sexual encounter is ambiguous because in the love-play the body loses its everyday way of being a tool. The gesture no longer expresses something definite – and thus it expresses everything; smiling and blushing do not simply mean kindness and diffidence, but a deprivation of meaning and a desparation of the body. It serves the purpose of letting the other become present; but at the same time it also conceals him: "*Anwesen*" remains here a *becoming* present.

The body is "used" to *do* something; it functions as a useful medium through which we realize something in the world. In our everyday life we understand expressions as phenomena which accompany our activities. In the sexual encounter the opposite is the case: *here we must understand the activities from the expression.* The whole body remains merely the medium for the expression of this one desire: to be all in all to each other. The body no longer receives its meaning from the world as the realm of all goal-oriented activities. It is necessary in the sexual encounter, and yet at the same time it is strictly speaking superfluous. That is why our hands are so alarmingly empty in amorousness, why we do not complete any of our sentences, why we stutter, blush, and smile. We do not know what to do, we become simultaneously warm and cold, experience emerging tension and at the same time a relaxation setting-in; the body still doubts because it is used to meaningful, that is goal-oriented activities. In amorousness the body can only produce a "strange feeling."

As soon as the lovers get close to each other and the sexual encounter begins, the body at once receives a new meaning; but this precisely confirms its ambiguous function. This meaning is: to serve exclusively as expressive medium; as a medium through which the person can become-present for the other as completely as possible for the sake of a dual mode of being. Thus the body must be "willing" to participate in a dual incarnation. From being my body it becomes a *part*-body and thus becomes meaningless in its distinctness from the other body. Body is always *my, your*, or *his* body; a medium through which a person exists. It is now attacked in this meaning. In the love-relationship one should be able to speak of *our* body. But there are always two of them. Thus in the expression: our bodies, two sorts of things are always included: the triumph of the love-relationship stressed by "*our*," as well as the failure and the continuity of the desire as found in the plural "*bodies*."

Desire implies the *participation* of the other, and thus also my participation. That is why we speak of *dual incarnation*. In the fact that the bodies of the partners always remain a duality we can find some motive for calling love "dupery." For

44. Cf. G. Gusdorf, *La découverte de soi* (Paris: Presses Universitaires de France, 1948), Part III, Chapter I.

what desire "intends" can never become really materialized. But our desire does not intend to materialize this *really*. Desire constitutes the love-play as the played realization of that which our desire "intends." It must remain incomplete if it is not to destroy its own meaning.

We wish now to consider the kiss and the caress from this point of view.

V

In a certain sense the kiss is meaningless and superfluous. No one has ever been able to indicate its *usefulness*. And yet one cannot declare the kiss unimportant: "Human love is distinguished from the being in heat of animals only by two divine functions: the caress and the kiss."[45] The kiss is an *expressive act* which belongs to the love-play.

We leave the *amorous* kiss out of consideration here; it does not fit in the context of the sexual encounter as we define it. Our interest is in the *passionate* kiss.

In the passionate kiss two mouths form one active organ. Each movement must be guided by the movements of the other – who in turn must be again guided by the other's mouth. In this sense the kiss is one activity by two persons. In love-play one does not really give a kiss, but it is enacted by both together. This possibility of dual activity founds the meaning of the kiss in the love-play. There is an important difference between the kiss and sexual unification in the narrower sense – setting aside the specific, genital tingling characteristic of the sexual organs – namely the fact that mouth and mouth are both of the same form to one another in this congruence. This alone is sufficient to show that the kiss has a completely different character than sexual intercourse: there is a symmetrical becoming-present of two people as equivalent partners in a common "activity." In a certain sense the kiss dissolves the sexual difference: the partners become present to one another in the kiss in the same way: just as he is present to her, so is she to him. But this is not yet all. The kiss *could* have been a dissolution of the sexual polarity if it had not precisely evoked the *differential* incarnation. For it has an erogenous influence and can even lead to orgasm. The fact alone that the kiss in the male can stimulate erection already signifies that it evokes the difference in bodilihood and confirms it. Within the perspective of the sexual encounter the kiss, therefore, receives a *provisional* character; true one can be taken up completely by the kiss and the encounter can fix itself for a while in the embracing, but sooner or later its erogenous working becomes overwhelming.

We do not wish to dwell here on moisture and warmth as primordial qualities of everything that lives. Both, mixed into one quality, are experienced in the kiss. That is why one says that in kissing the lovers drink one another's life, and even that they drink the kiss itself. From this there follows, also, an intimacy which no other form of touching can produce. In this respect we must see the passionate kiss as more "dangerous" than many "forbidden" contacts. The kiss reveals an inner

45. Louys, *Aphrodite*, p. 98, (English: p. 71).

warmth, a temperament in which the partner participates through the kiss. He *tastes* what *Bernis* merely *saw* in *Geneviève*:

> He thinks of this mysterious life which animates her, which makes her warm again as a sun, as an inner climate. Bernis does not say to himself that she is tender, nor that she is beautiful, but that she is warm. Warm like an animal: alive.[46]

The kiss is irreplaceable as a direct revelation of a warm inner self. And yet Proust observes that the kiss leaves us dissatisfied. Man, he says, is lacking a number of essential organs, and particularly one for kissing. That is why he uses his lips as a replacement for the missing organ. The lips must be satisfied with wandering on the surface.[47] What Proust did not understand is the fact that desire materializes itself in the sexual encounter as *desire*, that the entire body is the "organ" through which desire expresses itself, and that the kiss is one of the manners in which it manifests itself. It is precisely because the kiss confirms the desire that it remains attractive:

> Give me one thousand kisses and a hundred,
> And again some thousands and a hundred;
> Then once more a thousand and a hundred.
> When the number thus is many thousands
> Cancel it so that there be no traces
> Of the old account. Then start anew
> The counting till another count is due.[48]

In the encounter between mouth and mouth we become present for one another symmetrically *as if* the twofoldness of our bodies were not in the way of *our* unification – that is the unification of you and me as persons. In this act the mouth is taken up in the process in which the body gives new meaning. It does not function as an organ for the ingestion of food and for speech, but as an "expressive" organ. Its activity is an activity in function of the expression. This is also the way in which the poet puts it: "Lips which have kissed for the first time form a new mouth."[49]

VI

The fact that in the sexual encounter the body has another meaning than it has in the everydayness of our existence, can be discovered again in the *caress*. Here, too, we approach one another in a direct contact much more closely than we usually do.

46. A. de Saint-Exupéry, *Courrier sud* (Paris: Gallimard, 1929), p. 145. See for the meaning of the mouth: F.J.J. Buytendijk, *De vrouw* (Utrecht: Het Spectrum, 1951), p. 194. Cf. also: G. Bachelard, *L'eau et les rêves* (Paris: Corti, 1947), pp. 135-136 concerning "warm moisture."
47. M. Proust, *Le côté de Guermantes*, vol. II (Paris: Éditions de la Nouvelle Revue Française, 1920-1921), p. 232.
48. Catullus, quoted by C. Nyrop, *The Kiss* (Chicago: Allen, 1898), p. 54.
49. B. Aafjes, *In den beginne* (Amsterdam: Querido, 1949), p. 31.

"The caress does not want simple *contact*...: it is a *shaping*. The caress causes the other to be born as flesh for me and for herself." Or, in other words, "desire is an attempt to incarnate the other's body."

Sartre correctly says that caressing is not *taking hold of* a part of the other's body, nor is it *bringing* one's own hand actively *to* that part, but a "placing against it."[50] That is why he calls the caress the language of desire: sexual desire expresses itself as caress; and even more, it is the caress which first *materializes* desire.

Thus it is understandable that the transport of passion denies all goal-oriented dexterity to the hand. Just as was true of the mouth, so here, too, the hand loses its *instrumentality*, it now functions in an expressing and incarnating sense.

What does it mean that in the caress one makes oneself body with and for the other body? For Sartre correctly says: the being who desires is a consciousness which *makes itself body*. Herein we find the meaning of the expression we quoted earlier: desire has fallen into complicity with the body: "in sexual desire consciousness is, as it were made flesh";[51] and the same thing is expressed in the following passage from Jules Romains:[52]

> I do not believe that a real man could ever see two beautiful naked breasts of a young woman without being overcome by a stupor which is at the same time agonizing and delicious, which puts all movement of the mind in suspense, suppresses al concurrent ideas, simplifies the world amazingly by installing it, at its center, in a realm of bewilderment, these two sweet twin idols.

There is here a complete transmutation of the world and a complete embodiment of the body. If in the transport of passion there is consciousness-of, and not just unconsciousness, then it is exclusively a consciousness of *that* body and of mine, and at the same time a consciousness of orientation toward a unification of the two. Does this orientation mean "the lust of the flesh" or the tendency "to appropriate a body insofar as this appropriation reveals to me my body as flesh"?[53] Is the final goal of desire a total *carnation*? If indeed the central thesis is correct, namely that the body is essentially a medium through which the *person* exists, then Sartre's analysis falls short in an important respect, and even in the heart of his intention. The final goal of desire is not *carnation*, but *in*carnation. In the sexual contact the other is not taken *in* his body[54] but encountered *through* his body.

The mistake becomes clear the moment we grasp the meaning of the following statement: "*Desire is an attempt to strip the body of its movements as of its clothing and to make it exist as pure flesh*";[55] thus desire intends the appearance of

50. Sartre, *Being and Nothingness*, pp. 389-390.
51. *Ibid.*, pp. 387, 388.
52. J. Romains, *Le dieu des corps* (Paris: Gallimard, 1928), p. 123.
53. Sartre, *Being and Nothingness*, p. 388.
54. Cf. L. Binswanger, *Grundformen und Erkenntnis menschlichen Daseins* (Zurich: Niehans, 1942), Part I, Chapter 2.
55. Our italics.

of the flesh, *sarx* as "*pure contingency of presence*."[56] But this discovery of the other's body as motionless pure flesh is a form of transformation of the sexual relation which we cannot discuss now. Normally the relations are certainly not such. There is no question of *necrophilia*; there is a living body involved through which, under the caressing hand, the other is almost completely present – as a person, while the caressing one expresses himself toward the other through his hand.

The caress is meaningless if the one who is caressed betrays the caress through counter-movements: "These slender buttocks, firm as small apples, almost impalpable: the touch would not suffice to know them if once they are caressed they did not quiver in order, as it were, to play on their own."[57]

That is why we are in complete agreement with Buytendijk when he says that in amorousness the careful exploring in touching contacts and caresses is the genuine manner in which partners become present to one another: they discover in the touch the immediate, reciprocal presence and their being-with-one-another.[58] One could respond to this with Sartre that this reciprocity does not exclude the possibility that ultimately the issue is about the contingent presence of the flesh. But then one denies the value and meaning of our experience. For what can be the genuine meaning of this "fleshy body" if not that of being a medium for the incarnation of the other person? If we pursue Sartre's argument consistently, the height of desire should be capable of realizing itself in the caressing of a *dead* body. Suppose one caresses a dead man, eventually even kisses him on the mouth – what could this be if not a desperate attempt to call the absent one back again as a person through his motionless body?

All sexual stimulation in the narrower sense receives its meaning in the sexual encounter from the motive of the encounter: to be taken up by and with the other. This is why in the caress, too, the body is at the same time a barrier: "I certainly could take Albertine on my knees, hold her head in my hands; I could caress her, let my hands go over her for a long time; but as if I had handled a stone containing traces of salt from immemorial oceans, or the gleam of a star, I felt that I merely touched the closed envelope of a being who through her interiority, yielded to infinity."[59]

True, we find this experience in Proust in an intensified form – for Albertine does not take up his approaches; but this does not mean that the essence of it could not be present in each sexual encounter, at least as implicit *uncertainty*.

56. Sartre, *Ibid.*, p. 389.
57. P. Caminade, *Les amies des hommes*, p. 28. Cf. H. Kunz, *Die Aggressivität und die Zartlichkeit* (Bern: Francke, 1946), p. 82.
58. Buytendijk, *Begegnung*, p. 481.
59. M. Proust, *La prisonnière*, vol. II, p. 230.

VII

There is no doubt that clothing, too, can belong to a man himself; precisely through his clothing he can become present as this or that man; but this is always in the mode of a *persona*, a mask in which he expresses himself concealing himself. Clothing, by which we may understand here also the décolleté, the veil, the see-through dress, plays an important part in the love-play insofar as it suggests the nakedness of the body. If two people wish to come to a complete incarnation, there must arise a desire to meet one another in nakedness. Naked we see one another "just as we are." True, even then one can still hide himself, but one can never become present while still dressed in the way he can in the complete visibility of that which is mostly his own: his body. That is why Géraldy, alluding to the unification, says:

> Undress. Come fast.
> Let us take one another. The best means
> to explain oneself without being taken in,
> is to stretch out, body against body
> Don't pout. Take your dress off.
> Our bodies, they, will be in agreement.[60]

Géraldy means to say that in the embrace where the bodies fit on and into one another a mutual understanding will come about which cannot be spoiled by misunderstandings. The naked body has the expressive value of a medium through which the other can become present completely. The unification of our bodies is the greatest coming-together of which we are capable. But this is so only on the condition that the unification itself is understood as an expressive "activity," namely as that from which the sexualized body receives its meaning. Therefore we must now turn to the sexualization of the body.

VIII

There is a great difference between the phenomenal bodilihood of man and woman. This difference is important not only in its *facticity*, namely insofar as it defines man and woman biologically, but equally so in our experience. The "normal" man and woman feel themselves being attracted sexually to the *other* sex as this manifests itself bodily (exceptions to this cannot be discussed here). This fact, which is so "obvious" that we are prone to overlook it, contains nonetheless a meaning, namely the differential meaning of masculine and feminine bodilihood. In the body of the man we see first of all the useful instrument. Muscular, calloused, marked with scars from his work, he is the one who *works* with his body. The feminine body on the other hand has a value in itself, is not judged instrumentally, but "aesthetically." It is not important for our present purposes whether these relationships are sketched in an "objectively" correct way. It is even quite certain that they are not.

60. P. Géraldy, *Toi et moi* (Milano: Editoriale italiana, 1952), pp. 81-82.

For "objectively" a woman works as much as a man. And yet the same work incorporated into another life-style, has a different experiential value. And we are asking here at the moment precisely for the experiential value which the body has for the other. The masculine body stands under the sign of instrumentality: he appears as *homo faber*. And this is so much the characteristic which dominates his mode of existence that his sexual organ, too, appears in this light. We wish to devote further consideration to this part of his body in that it, pre-eminently, is the male sexual characteristic, that part of the body in which the sexualization of the masculine body is invested for the greater part. For that reason it must be possible to understand the meaning structure of the masculine sexuality from the meaning of the penis in the sexual incarnation.

In theories on sexuality the penis is usually considered an instrument, and more precisely an aggressive, penetrating instrument which, in this quality, characterizes masculine sexuality completely. *This conception is incorrect*. True, the man himself, too, is often inclined to conceive of his sexual organ in this way, but this is so because his whole body is an instrument for him. If one part of his body did not have this meaning, what then could be its meaning? In other words, in any account of the body the sexual organ is integrated into the general scheme from which the body itself is understood. But the penis can in no way be justified in this way. The masculine sexual organ is, on the contrary, that which prevents a man from existing wholly as *homo faber*, at least insofar as he accepts this "organ" in its proper meaning: It is an *expressive* "organ," and more precisely the only "organ" which *primarly has an expressive function* (we will not speak here of its biological functions: excretion, secretion, because from a *psychological* point of view these functions do not first and foremost determine its real meaning). *Psychologically* we understand by "*organ*" a part of the body that has as its goal a certain achievement whose meaning is to be found outside the organism: for the hand manipulating in the broadest sense of the term, for the eye, seeing, for the ear, hearing, etc. All of these are activities or achievements – acts, if you wish – which are motivated by the scene and result in a change of the scene. We distinguish two groups of organs here: the executing organs or organs of action, and the receiving organs or senses. For instance, my manipulations are actions; they are motivated by objects with which I wish to do something and with which I do something. My seeing is a receiving function, the eye is a sense organ; in seeing I see something which attracted my attention and which, through my seeing, I place at the center of my visual field; in hearing, I hear something, etc. Both groups of organs serve an active relationship bebetween person and world. We understand the organ from its *function* – that is, an indivisible whole of changes which is meaningfully related to something which lies outside the organism.[61] On the other hand, expression or expressive motion is neither an action, nor a reception – and therefore not a function in the sense defined here. "The action is related to its end or goal; the expression *which has a meaning in itself*, is related to the mode of our being-in-the-world."[62] If we define

61. Buytendijk, *Algemene theorie*, p. 25.
62. *Ibid.*, p. 134. (Our italics.)

an organ as organ through its function, then an expressive "organ" is not an organ, not an *organon*, that is not a tool or a sense. In due time the organs, too, can express something, as it were secondarily, something we think is contained in the interiority of the other: an emotion, a feeling. The behavioral pattern is then understood as an expression and the expression in turn is understood as an epi-phenomenon, as a clarification of the action. The hand is an *organ for work*; it will never happen that we will wish to see its meaning primarily in its expressive possibilities. In our everyday life we consider these possibilities to be identical.

We must now ask the question of whether the penis, psychologically, has the function of an organ in this sense. Faithful to the psychoanalytic tradition, Federn has placed the masculine libido as active opposite the passive feminine libido and in so doing has compared the male sexual organ to the *hand*. Or rather the other way around: the pre-eminently active and aggressive organ is the penis; the hand derives its value as a sexual symbol from this.[63] In a certain sense this idea seems to be quite obvious. For in what way does an *erection* address itself to us? Sartre gives a brief indication with which, generally speaking, we can agree: "... that thing begins to move like an animal, it gets hard, it frightens me, and when it is stiff and sticks straight up into the air it is brutal...."[64] Phenomenally, in the erection a spasmatic force manifests itself: the phallus is a loaded explosive in which all the tensions of the male body are centralized. In its not erect state the penis is certainly not the perceptible center of gravity of a man's nude body. True, the eye does not completely by-pass it and is even stopped in the flowing movement of its observation by it, but then nevertheless the eye neglects this obstacle for the sake of the whole. Look at a classical Greek statue: the penis is an accidental *appendix*, an innocent part of a body whose force and muscularity we admire, balance and posture, action and expression. But in the erection all of these qualities flow toward this one part of the body. It destroys the aesthetic value of the whole in that it commands too much attention, and furthermore sexualizes the body in an unambiguous manner. The figure's center of gravity shifts towards this demonstrative center, this erected and tense member behind which the man stands in his entire force. It is the undeniable sign of *potency* in every meaning of the word. If it has an aesthetic value, then it will be found in its frightening and impressive beauty: "this kind of impatient, terrible beauty."[65] One would like to speak of the *mana*, that mysterious primordial force, which perhaps could also explain the phallus-cult and the worship of the *lingam* in antiquity.[66] The phallus-symbol not only stimulated sexual potency, but also conquered all kinds of evils, increased the fertility, etc. The male organ was exalted to a god-status because of the procreating, impregnating, and even creative force which manifested itself in it; it pre-eminently incorporates the active

63. P. Federn, "Beiträge zur Analyse des Sadismus und Masochismus," in *Internationale Zeitschrift für ärtzliche Psychoanalyse*, 1(1913), pp. 38ff.
64. Sartre, *Le mur* (Paris: Gallimard, 1939), p. 97.
65. J. Romains, *Le dieu des corps*, p. 136.
66. Cf. J. Rosenbaum, *Geschichte der Lustseuche im Altertum* (Berlin: Krayn, 1904); also: H. Flournoy, "Çiva Androgyne," in *Archives de Psychologie*, 18(1923).

principle. *That is so, at least when the penis appears in isolation.* In the *phallus* and the *lingam* the penis is almost spiritualized: it becomes a symbol of primordially creative force; a medium in which, according to the experience of ancient peoples, this primordial force becomes present in the way a god becomes present in the temple. And this is so not only because of the penis's *genital* value but especially because of its *generative force* which as you know in ancient times was especially attributed to the man, a view which is in complete agreement with the active, creative, masculine life-style. This is an illustration of how the essence of the man as generator, procreator, is embodied in the phallus, but on the condition that we isolate it and at the same time conceive of it as the embodiment of masculinity.

The description of the phenomenal quality of the erection just given rests on a mode of experience which is found very often and *sees the penis in function of the instrumental essence of the masculine body*. This seems to be obvious: for the penis is that which defines a man as man. Biologically this is correct, *but psychologically it is by no means correct*. Psychologically we understand a man as man primarily from his masculine form of life, which is an equally possible life-style for the biological woman: a style of life which is defined by work, through the world of work. "The world of work is one of resistances which form our existence through the interchange of tension and relaxation, which constitute our existence as solid, hard, and strong, as definitely grasping, as an existence of being-able-to, ought, and willing, of conquering and courageously being."[67] From this masculine world of work we understand the masculine body as an instrument, as *the* instrument through which it can engage in work. The parts are involved in this constitution of the whole's meaning, and thus we must relate to that constitution, too, that part of the body which *bodily* stamps him as a man. But if, indeed, the activity of the sexual organ is oriented toward this world, and in harmony with the "spirit" which the phallus-cult attributes to it, even the most masculine of all activities, *why then is it that sexual activity is so little integrated into the genuine life-style of the man*? Why then is it that in the memory of man, man has always pretended that in the innermost depth of his essence – that is as *homo faber* and "spirit" – sexuality is *alien* to him? Why is it that he has placed the "blame" for sexuality on the woman? The answer must be: this is because his sexual activity is *not* "work," the penis is *not* an instrument, not an organ for work in the sense just defined.

Let us compare the penis with the hand. The hand which is the organ for grasping, holding, touching, feeling, giving, receiving, doing, and handling, in a word the organ for manipulating, is in harmony with this function – a muscular, mobile, articulated and differentiated organ, with infinite variations as far as possibilities of movement are concerned. The penis does not show any of these characteristics. One cannot *do* anything with it, at least not instrumentally. Even for stimulation of the female sexual organ it is the hand which is infinitely more suited. The penis is *not* muscular, and this objective characteristic already has an important meaning: for as far as the masculine, "industrious" relation toward the world is concerned the man has transverse muscles at his disposal. The penis is not articulated, nor mobile, at

67. Buytendijk, *De vrouw*, p. 291.

least not through the musculature proper to this organ. It is not at all a manipulating organ; at the most it is a manipulatable organ; thus it is not active, but passive. Furthermore we see also that our hand, this more or less independent father of all tools, has its own apparatus at its disposal through which it can move: the arm. In order to shift or move the penis, on the contrary, a man's entire body must put itself into its service. It is this "service" which makes the masculine body clumsy, awkward and maladjusted in the embrace: this is obviously not the real task from which the man has implicitly understood his own body. A comparison to the natural suppleness of the woman in the embrace turns out to the man's disadvantage. Her body is already attuned to expression; the man must first *transform* his. When he succeeds in doing so he loses his clumsiness; then his entire body becomes what his penis is already: expressive "organ." Then there can no longer be a question of "service": his entire body has become penis.[68]

Whereas manual movement is pre-eminently voluntary, guided by "intentions" to execute "plans," the erection which gives the sexual organ its meaning is totally involuntary. The hand is *at our disposal*; the penis is not. The hand obeys, the penis does not. A manual movement is something we can do or not do, at the most the erection – to a greater or lesser extent – can be *suppressed* or smothered by thinking about something else. It expresses a mode-of-being in the situation.

If there is "aggression" in the sexual organ – which is not *its* movement – such as is true, for instance, in defloration, then this aggression must be attributed to the masculine body standing behind it in an untransformed way. Even the *immission* does not always take place spontaneously, but usually with manual assistance. As an instrument the sexual organ is completely unfit. If the immission occurs spontaneously in the love-play it is experienced as a special gift: for there it first becomes manifest how much any manual help and any instrumentality *disturbs* the real sexual "activity."

The sensitivity inherent in the hand is passed over in the manipulation; in touching and handling something we do not feel our own hand, only the touched and handled thing. It is only in the sexual revalidation of the body that the sensitivity proper to the hand is first materialized as erotic sensitivity, as we have seen earlier. The sexual organ, on the other hand, is unable to pass over itself in this way.

Thus the comparison of the penis to the hand falls in every respect. The erection is the direct expression of the manner in which a man at the height of his desire becomes present in his ambiguity: the erection breaks through the instrumentality of his body. That is why he no longer knows what to do with it, with this undeniable expression of an enormous, but helpless *tenderness*.

But this meaning suggests itself to us only if we see the erection in function of the total situation and no longer as an isolated happening. The erection expresses the sexual incarnation: that is to say the willingness to be all in all to the other and in this willingness the disappearance of all plans and intentions, of the entire world as a field of goal-oriented activities. Hence this "consciousness made fleshy." This willingness is in conflict with the general type of masculine existence. For the erec-

68. This expression is from van den Berg (personal communication).

tion does *not* mean an activity "which flows from an intentional act in which consciousness is oriented toward a proposed goal which is independent in regard to the activity." The erection does *not* constitute the masculine form of existence taken "as a being involved in the projects of his labor and in what is offered to him as material in the actuality of his world."[69] In the erection the man's existence constitutes itself as *not* autonomous, *not* prospective, *not* aggressive, but rather as expansive. For here it is dependent upon the other, and even commits itself to the other; it does not look ahead, but engages itself in this one and only, suspended moment; it is not aggressive, but *tender*, oriented toward the experience of bodily contact in which the love-desire of both partners manifests itself, and thus toward a dual incarnation.[70] Finally his mode of existence is expansive in that here the man goes beyond himself, transcends himself, but not in that centralized manner of all goal-oriented acts which focus only on one point, but in the fading and undetermined way of the emotions.

The penis is not an organ for action. On the other hand it seems clear without further explanation that it does function as a sense. If we conceive of the penis as a sense-organ we get what is called "sexual lust." It is wise to lable this specific experience not as sexual but as genital lust. The possibility of experiencing this sensation in its specificity and for its own sake, founds a form of sexuality which is more a relationship to one's own body than to the other. The man in particular is often blamed for the fact that sexual intercourse is not *intercourse* for him, but an autistic satisfaction of his own genital lust. Phenomenologically seen, this attitude which degrades intercourse to an "each-for-himself," refers to a secondary form. The woman is not a partner here, but a "pleasure machine."[71] In this case the body has an instrumental meaning for the man, sexual intercourse becomes "onanism-for-two," in which the woman usually never gets her own share." Sartre describes the "relationship" between Antoine Roquentin and Françoise in this way: "we make love at par.... Each one for himself." The absence of an encounter in the pregnant sense is confirmed in the scene by the fact that Françoise quietly speaks of a kind of wine, and in passing asks whether Antoine would object to her keeping on her stockings.[72] There is a business-like relationship, no more.

It is not without reason that we speak of onanism-for-two. For what we observed earlier about masturbation, is characteristic here, too: *the partner is missing.* He or she may be present "objectively," but he does not function as someone who participates, at least not in an authentic way. Does this mean that auto-erotic sexuality which turns itself back toward one's own body is primarily founded in genital lust? By no means. We may think here of the pictures often used by the masturbator, or of fantasies in which *someone else* is always evoked. Sexuality

69. Buytendijk, *De vrouw*, p. 290.

70. Cf. Kunz, *op. cit.*, p. 77.

71. G. Blin, "Le sadisme de Baudelaire," in *La profondeur et le rhythme* (Paris: Gallimard, 1948), p. 252. Cf. D.A.M. Marquis de Sade, *Justine, ou Les malheurs de la vertu* (Paris: Pauvert, 1950), p. 191.

72. Sartre, *La nausée* (Paris: Gallimard, 1938), pp. 18-19. Cf. the scene in *Le mur*, pp. 74-75.

directs itself primarily toward the other who then, however, need not be accepted as a partner, and even can be *denied* as partner.

If we dwell for one more moment on genital lust itself, we must note that the penis, if it is a sense organ, is a remarkable kind of sense organ. Its awareness does not contribute to our knowledge of the surrounding world, not even to our "knowledge" of our own body, such as proprioceptive awarenesses. The lust-feeling is a sensation in the strict sense, a bodily feeling, an experience of one's own ecstatic bodilihood. One cannot take distance from it. Bloch, who certainly did not pursue a phenomenology of sexuality, already called orgasm a losing of consciousness, a giving-up and "dissolving" of one's own personality.[73] Completely in harmony with this Buytendijk writes that the lust-feeling removes the subject-object opposition. The person forgets himself by going up in a sphere of uninhibited "fullness of life," joy, and bliss. The lust-feeling is not experienced as such, but it fills us and moves us to ecstasies.[74] No wonder the lust-feeling eludes the person the moment he pays attention to it. It is then reduced to a simple pleasurable experience from which disappears just that which was precisely at stake. If the orgasm is the summit of this incarnation in which the person as self-consciousness coincides with his body, then there is no leeway left for a relationship with one's own body. That is why Bachelard says: "It is impossible to escape from this dialectic: to have a conscious awareness of burning is tantamount to cooling-off; to experience an intensity is to diminish it: one must be intense without knowing it."[75] *But if this is so then genital lust in the narrow sense is a residue.* Lust which is completely aware of itself cannot be related to the person who is taken up by it. Then orgasm is not a private affair, but a communion, an ecstatic transcendence of one's own limits, an experience of the *other* in one's *own* body. The saying "omne animal post coitum tristis est" (every animal is sad after intercourse) holds true only if I withdraw from the communion, break through the partnership by turning back toward my own body. By doing so the orgasm really becomes a failure. It is no longer an orgasm, but merely a pleasant awareness and thus a failure of that toward which it intended. This motivates the "tristis esse" (being-sad).

When we call the orgasm a communion, this does not imply that the penis is a sense organ in the common sense of the term. A sense organ is a communicative organ, attuned to certain stimuli which are passed over toward the things and the qualities which are perceived. We perceive nothing through the penis. Orgasm is not a perception. There is no "sense-organ-of-lust." Orgasm is an expansive surrender which is materialized in pure bodily joy. Through orgasm the emotion which is materialized is expressed by it, becomes flesh. The person has renounced his past and future and all they promise, and at this moment coincides with his body. That is why orgasm is a completely unambiguous expression: there is nothing behind it. In orgasm a total, momentary becoming-present expresses itself before the other and is oriented toward the other. It is in this sense that orgasm is the completion, but at the same time the end of the dual incarnation. "Our bodies, they will agree."

73. I. Bloch, *Das Sexualleben unserer Zeit* (Berlin: Marcus, 1919), p. 45.
74. F.J.J. Buytendijk, *Über den Schmerz* (Bern: Huber, 1948), pp. 20-21.
75. G. Bachelard, *La psychanalyse du feu* (Paris: Corti, 1949), p. 218.

The penis is neither a tool nor a sense organ, and thus not an organ, either, in the sense defined above.

We have dwelled on this point because our conception constitutes an attack on those theories which wish to derive our sexuality from a property of the sexual "organs," namely the fact that they procure a typical "lust-quality." Often an aggressive component is added to this. However human existence cannot be built up through the execution of the "objective properties" of man's body. From a psychological point of view we may deny the part of genital lust-feeling and aggressiveness as *founders* of our sexual life. In its very origin sexuality is *relation toward the other*. It cannot be denied, however, that there are transformations of the original sexual relationship. The fact that these transformations eventually in fact occur more than the "genuine" form of sexual relationship does not invalidate our analysis. If this indeed is the case, it merely stresses the fact that the realization of the sexual relationship in its genuine form is a *task* in which one can fail. The nature of this task is predelineated by the manner in which our sexuality reveals itself to us through the sexual organ. The sexual life of this or that concrete person is just as little a simple natural happening which develops according to its own and unchangeable laws in this or in that way, as is his entire existence. His existence, and equally its sexual aspect, is an *ethical* existence: it unfolds in the realization of *values*. Well the *value* of sexuality becomes already manifest to us in the erection, as we have just analyzed it: *incarnation because of the other and because of ourselves* (we are unable to discuss here the fact that homosexuality is not excluded in this view; we are just mentioning it in order to avoid any misunderstanding of our conception). That is why no *psychological* sexuology is possible without a reflection on love as a form of being-together which manifests itself in sexuality in a definitely original and irreducible way. We must refrain from such a reflection because it would require more space than is available to us here.

There certainly is an aggressive variant of sexuality. We encounter this form in Sartre: an attempt to integrate sexuality, too, into an exaggerated masculine form of existence. We have at the same time a sexuality which strives for a goal within one's own body: a hedonic genitality which tries to reduce itself to a search for lust. But our goal here is not to pay attention to the various forms of sexuality, but to the meaning of the human body in the sexual encounter. We have found that the meaning of the sexualized, masculine body consists in the fact that it expresses sexual desire, not primarily as a desire for lust, but as a tender desire for the other; it expresses the willingness to let one's own existence as a person come together with that of the other. This "emasculation" of his body – one would almost be inclined to speak of a feminization – makes the manner in which the man becomes present in the sexual encounter principally an ambiguous one.

IX

If it is true that the body of the man is first and foremost an instrument and that its expressive qualities are secondary, then the opposite holds true for the feminine body, at least as long as we are speaking schematically. The latter is primarily an expressive medium – that is, it is experienced and judged in this way. Strictly speaking in so doing one says little else except that the distance the person maintains in regard to his own body is "smaller" in the woman than in the man. This does not mean that she is *more* body in an alleged opposition to the man who would be more "spirit," but merely that she is *more directly bodily*. Simone de Beauvoir expresses the same thing by observing that in a special way the body of a woman is *hysterical* in the sense that there is often no distance at all between "contents of consciousness" (particularly in "emotional states") and their bodily expression.[76] On the other hand, perhaps feminine "intuition" rests on the fact that the woman less than the man searches for something *behind* the expressions and so is more able to directly understand subtly shaded expressive movements in connection with the situation. At any rate, there is little doubt about her greater vegetative lability because of which she experiences all that is bodily in each situation in a different way from a man.[77] The question of whether the relations are perhaps just the other way around, and the woman precisely because of the fact that she is in the situation in a feminine way is vegetatively more labile than a man, is not important here. We combine both these ways of looking at things in the observation that a woman is more directly bodily. The *anthropological* characteristic of her bodilihood seen from a psychological point of view, has the remarkable consequence that a woman seems to possess more interiority than a man – as we shall see in a moment.

What sexual meaning does the feminine body have? This question must be specified more precisely. The feminine appearance "shows the intended presentation of what is soft and thus rounded-off, smooth, of what can be caressed, of a profusion which is gratuitous and thus a grace. The feminine body makes perceptible the act of her susceptible surrender, that is, of her expression of tenderness and her need for tenderness. Although it is true that these acts also give human feminine form to the execution as well as to the experience of a sexual function, it is completely incorrect to think that the bodily appearance of the woman means sexual attraction."[78] Through what, then, does this appearance become sexually attractive? Or better, through what is the appearance *not always* attractive? For it has been said from early times on that the body of the woman is a symbol of happiness and mutual satisfaction,[79] that her body has been created in order to be embraced. Pierre Louys lets Chrysis use this view to justify feminine homosexuality and at the same time condemn the masculine homosexuality: from head to foot she, and she alone, is made for love and wonderfully so. She alone knows how to love. She alone knows how to be loved. From this it follows that when a love pair involves two women, it is perfect; if it implies only one woman then it is merely half as good; if there is no woman involved, then it is sheer nonsense.[80] It is a fact that a "normal"

76. S. de Beauvoir, *Le deuxième sexe*, vol. II (Paris: Gallimard, 1949), p. 156.
77. Buytendijk, *De vrouw*, p. 113.
78. *Ibid.*, p. 195.
79. C.A. Scott, "Sex and Art," in *American Journal of Psychology*, 7(1896), p. 206.
80. Louys, *Aphrodite*, p. 97. (English: p. 70).

man does not feel attracted by another man whereas a woman remains sexually sensitive to the look of another woman, particularly to the face and breasts without being homosexual for that reason.[81]

It is her expressive movements, or the feminine movements insofar as they are conceived of as such, which as the fulfillment of the expressive essence of her body, sexualize the feminine body. If her appearance is not always attractive, then this is true insofar as she transcends the pre-given meaning of her body through an intentional, goal-oriented concern with things, or has transcended it definitively through her orientation toward a loftier world. Since we have rejected placing sexuality and genitality on a par, this utterance cannot mean a "degradation" of the feminine body into an incarnated genitality. But this enunciation does require further specification.

Here we may quote and agree with Hesnard who, in contradistinction to Sartre, for whom "motionlessness" seems to be an ideal, says that what is important is the affective game which appears here, the physiognomic and mimic expression, the life of the human feelings as they come to the fore. The most attractive feminine figure if struck by motionlessness, would, as far as desire is concerned, already contain a germ of death..." it is particularly movement which evokes sexual desire.[82] This is especially true for the expressive movements. But not all expressive movements are erotic. That is why Hesnard speaks of specific, erotic indices in the expressive movements, although it is true that he does not specify them in detail. We can perhaps find this specificity in the *luxuriousness* of certain expressions. In the everydayness of our existence we are inclined to understand expression as a clarifying epi-phenomenon of our activities; we have already noted this before. Since this activity as this or that specific activity is understood from the situation, we also understand expression from the meaning of the situation. The expressive movement we are referring to by the provisional term "luxuriousness" must be, then, in contradistinction to the "normal" expressive movements, a movement which *itself* constitutes a situation and thus reverses the "normal" relationships. Such an expression breaks through the canvas of teleology on which situations and activities usually develop, and thus forms the "open field" of the immediate Eros about which we spoke before. This means nothing except that the horizon of the encounter changes. The extreme example is the smile of the *Mona Lisa* which refuses any interpretation. This smile *is* simply there; it constitutes an a-practical situation in which one cannot "do anything": it disarms, it does not take a position and does not allow one to take one; it transforms the masculine world into an expressive world which has its value in itself and thus deprives the masculine body as instrument of its meaning, because of the fact that its horizon of resistance-to-be-overcome falls away. This is true not only for this smile, but also for countless "useless" and "aimless" expressive movements which seem to be performed merely for their own sake. Everyone knows from his own experience these charming and glamourous, "useless" feminine peculiarities found in her posture and movement, her walk,

81. A. Hesnard, *Traité de sexologie* (Paris: Payot, 1933), p. 283.
82. *Ibid.*, p. 284.

and the way in which she combs her hair. Everyone knows this mysterious gesture filled with itself which, in the eyes of a man, stamps her as a "woman" in the very pregnant sense of the term. We must not conceive of this mysterious gesture as an aesthetic phenomenon. It is extremely difficult to distinguish between the "really feminine" and the aesthetic. Without going into this question here we simply wish to note that this distinction can be found in the *personality* of the erotic gesture. In the aesthetic phenomenon gesture exists as the incorporation of beauty *as such* freed from the personality of human existence: "*beauty itself*" becomes manifest in the gesture. In the erotic gesture the feminine *person* becomes manifest. In this way we further determine the "useless," erotic expression as an expression through which the woman as a feminine person becomes present, and in so doing fulfills the pre-given erotic meaning of her body.

But this formulation, too, is still unclear. What is it that expresses itself in the erotic expression? What is the erotic meaning of the feminine body? We let ourselves be put on the right road by the poets who celebrated the feminine body in a way in which the masculine body has never been celebrated:

> It is like a flower of crimson, full of honey and of perfumes.
> It is like a hydra of the sea, living and soft, open at night.
> It is the humid grotto, the shelter always warm, *the Asylum where the man rests from his march toward death.*[83]

A comparison to the scene is found, albeit more sporadically, in connection with the masculine body – just think of the Song of Songs. Nowhere, however, do we find for the masculine body the symbolism: resting-place on the road toward death. This symbolism contains nothing but what we have already seen: the sexual relation calls the man back from his involvement in an existence which is a proceeding, autonomous, aggressive, and prospective existence conquering resistances. This relation urges him to come to a standstill. For what? In *Le vagabond sentimental* a love scene between Baccio and Erigone is enacted which is described in a manner which is important in more than one respect:[84]

> I am traveling, he said, I am making a long journey in a country of aromatic substances. Each part of your body has its own perfume and I am going from one to the other, like a merchant of essences and gums.... Wait for me, I shall be right back....
>
> ... When he returned laden with fragrances and rich as a merchant of Baghdad, she received him with the raptures of a woman who had been neglected for a long time. And from this playful journey they proceeded little by little to the most lively ardours, searching in themselves, at the bottom of their flesh, through their sighs and murmurings, beyond all consciousness, in the ideal

83. Louys, *op. cit.*, p. 26. (English: p. 19.)
84. A. T'Serstevens, *Le vagabond sentimental* (Paris: Michel, 1923), p. 135. (Our italics.) English: *The Sentimental Vagabond*, trans. W. Chambers, (New York: Farrar and Rinehart, 1930), pp. 152-153.

> light which their passion kindled, for *the great mystery of life, the origin and the end.*

Baccio is so much of a man that even in the scene of love he has to take a *journey* and acquire possessions. That is why Erigone says when he begins his journey: "What are you doing there? Are you forgetting me?" But he does not forget her; he returns heavy with treasures in order to experience with her the "great mystery of life, of origin and end." It is this mystery which is manifested to him by his retiring into the feminine body.

To conclude these quotations we quote a passage of Jean Legrand:[85]

> Inattention could play a part here for it accompanied the *poetic* liberation of the *natural* world which would have been limited without it. To grasp in a bosom the masses of waters and winds, in the flanks the caves and underground mines, to find in the muscles which *possess* sometimes the curves of rest, sometimes the straight lines of force, the primordial consecration of the fruits, the most refined products of the globe ... to act, to make love a universal gesture from which nothing could ever escape, that is, after having grafted one sex upon the other in complete attention, to carry a total consciousness of a point over to a surface – the open cover of the body on the central superimposed layers – and to create oneself in that state of the world and of a mirror without depth.

Resting-place, revelation, and recreation; richness of the scene that is "just merely" there in a demonstrative and luxurious way; do these qualities manifest themselves in the everyday erotic gesture of any arbitrary woman? Are these qualities indeed the qualities of her bodilihood?

It is the bodily appearance which expresses what is feminine. For instance the breast which, independently of her motherhood, is present in the mature woman, "is the exteriorized softness, the 'feminine heart' in its motherly kind-heartedness." Her arm is "rounder, softer, the hand is narrower and smaller, less instrumental as far as outward appearance is concerned, less visibly adapted to the forceful grasp of work."[86]

It is not so much motherliness taken as possible or factual motherhood which sexualizes the body of the woman. The expressive value of the breast does not refer for any man to *this* motherhood, and it certainly does not mean a fixation to the feeding breast. It may be true that the Western world no longer knows anything of the secrets of the "magna mater"; it still speaks to us through the breast of the woman: "that great mystery of life." This is a life which has not yet fixated itself through decisions to go straight forward toward the goal, determined and certain of itself; it is still *open possibility* which is self-sufficient: pure luxury. Thus pure luxuriousness manifests itself in the breast which is always there without always being *useful.*

85. J. Lyrand, *Aurette et Jacques* (Paris: Corti, 1947), pp. 225-226.
86. Buytendijk, *De vrouw*, p. 188.

We understand as open possibility, too, youthfulness, life "without pronounced grasp on the world, without demonstration of expansion as a mode of being which transcends itself." There is further the relaxedness of her greater symmetry: "intended suspension of the intentionality which is directed towards the world." The "genuine" feminine asymmetry is furthermore not ambiguous as is that of the man, but harmonious and unison: "it is a rhythmic inner relationship of the parts in regard to one another, a melodic play of lines."[87]

This "natural" expressivity of the feminine body *can* be transcended. She need not constitute herself as a person who *is* in the way in which her bodilihood signifies her. As a person she can become present through this body as someone who denies the "natural" indications of this body, but also as one who affirms and fulfills them. This fulfillment of the "proper meaning" of her body is enacted in the erotic gesture – and this is true even for the most angular woman: the woman as a *person* fulfills the meaning which is predelineated in her body in its "useless" tenderness, in the relaxed movement of an arm, in the purposeless smile, in the "superfluous" diffidence, the shy blush, the waiting attitude, the unintentional showing and marking of her bodily forms in posture, walk, and way of dressing, and thus in every expression which shows the openness of an existence without tension and effort. That is why her body is not *per se* attractive, but only when in the eyes of the other she becomes present through her body as a person who promises the revelation of the mystery by re-laxing the man's existence, through which the man will be more able to understand her in a "recreated world." It seemed to me, Proust says, that the beauty of the trees was hers, and that the soul of these horizons, of the village of Roussainville, of the books I read that year, would be revealed to me by her kiss: "my desire no longer had limits."[89]

In her erotic expression the woman fulfills the immanent meaning of her body as a provisional outline of her femininely-being-a-person.[89] That is why the meaning of her body in the sexual encounter is something other than the meaning of the masculine body. The body of a man *transforms* its meaning from instrument to expressive medium. The body of the woman *fulfills* its immanent meaning by revealing itself as expressive medium. What the man – and she herself too – discover in this body is for instance, that she does not have a specific "organ" for her sexual incarnation. In this regard the vagina cannot be compared to the penis. Even sexuality is centralized in the man: he has a part of his body which unmistakably expresses his sexual incarnation. The body of the woman in its entirety is sexual "organ." Her desire spreads through her whole body; it does not especially incarnate itself just where the male "organ" will encounter her. Her entire body is one erogenous zone although it is true that one can indicate special erogenous places on it.

We must dwell for a moment on the concept "erogenous zone." Does this mean that there are determinate parts of the body which are particularly appropriate for

87. *Ibid.*, pp. 201, 217, 218.
88. M. Proust, *Du côté de chez Swann*, vol. I, p. 145. (Our italics.)
89. M. Merleau-Ponty, *Phenomenology of Perception*, pp. 230-232.

evoking sexual desires? By no means. The erogenous zone is a place where sexual desire materializes itself, where I particularly incarnate myself under the touching hand of the other. There is no reason whatsoever why this touch *per se* should be erogenous. Erogeneity is determined by the encounter, by the acceptance of the manner in which the other becomes present in this encounter as the one who caresses me. Erogeneity is not an objective property of certain parts of the body. The udder of a cow does not function as an erogenous zone. But why then does the breast of a woman function in this way? Why does the earlobe do so, for as is generally known this can be extremely erogenous? Why the skin and the lips? Erogenous zones are different for different people. This is not so because of some differences of their physiological constitution, but through a "decision" that they are prominent points of the sexualized body for them. The caressing hand of the other is necessary to make them erogenous, just as tickling gives us its specific sensation only when *someone else* tickles us and even then only when the one who is tickled agrees to being tickled.

The woman's body is erogenous from head to toe. Attractiveness, too, implies the entire body. The most beautiful figure can be spoiled by a face "which does not fit," and even by the eyes alone. The cult of "perfect bodily forms," of feminine beauty in general, is related to the entire body because it is, in its entirety, her sexual "organ," – in any case it should be so. What is more determinative for the overall erotic impression, the hair, the face, the breasts, the abdomen, the genital "triangle," the legs? Even the fingers, the ankles, the line of her neck play their part in the evaluation of her sexual attractiveness. It is the whole that must be "perfect" through the "rhythmic, inner relation of the parts to one another, a melodic game of lines." And the manner in which she behaves with her body in the embrace, too, is rhythmic and melodic, graceful and supple. She does not have any trouble with this encounter.

Whereas her sexual incarnation implies the fulfillment of her beauty, an appeal to her beauty is, on the other hand, sometimes already sufficient to make her capitulate. It is an old adage that one must praise her beauty in order to win the woman. When the lover says: how beautiful you are, he invites her to accept the quality of her body as *hers*. Properly speaking he even dedicates her own body to her: he holds it transparent to the person who, through her body, becomes present in her beauty. By accepting this dedication she fulfills the beauty of her body – and is "lost."

The woman incarnates herself sexually by *becoming* what her body already expresses. She *disposes* of all goal-oriented activities. That is why she appears to the man as the one who capitulates, surrenders, and delivers herself. Her appeal to his protection is contained in her surrender. She gives the man a body which does not protect itself, but indeed is vulnerable, and in the beginning even sealed. The hymen is indeed a "resistance" which invites the man to behave as a man – although it is in contradiction with the intention which motivates his sexual "behavior." It is here that the myth of the "original" aggressiveness of the male sexuality originates which later all too often materializes itself in an ensemble of misunderstandings. If for the woman the penis has the meaning of being an instrument for taking-possession-of, if the man in turn believes that he has to *take* the woman, in which case both are motivated by one aspect of the mode of appearance of the other, then future sexual encounters will be easily fixated in the scheme of "taking and being taken." That is why the defloration is such a tender point. It is often in this first complete encounter that the decision concerning their being-together is made. How much the woman is expecting tenderness, and not aggressiveness, respect for her personality and not genitality is clear from the fact that she is hurt as a *person* when the "flower" is not plucked but crushed. She can then turn away as a *person, close herself off*; and in view of the fact that she is more directly bodily than a man, she expresses this through vaginism, frigidity, and perhaps even disgust.

When we say that a woman "gives herself," it does not mean that she offers herself to be taken, but that she is willing to *receive*. Giving and receiving are one here, just as hospitable receiving is really giving, namely a giving of oneself.[90] She receives the man and gives him a home.

The relationship between femininity and home is not just a traditional relationship established through her function as a housewife. Home means more than just the material shelter. It is not without reason that one speaks of "the home of love" (*Heimat der Liebe*). In view of the fact that the essence and meaning of this "*Heimat*" have been explained by Binswanger in a way as excellent as it is "unbodily," it may suffice here to refer to his work.[91] We shall limit ourselves to the meaning of the body and summarize our explanation thus far as follows:

In the sexual encounter the body of the man expresses the willingness to go into the other; the body of the woman expresses the willingness to receive the other. The unfortunate opposition "active versus passive" has its origin in this relationship.

The woman fulfills the immanent meaning of her body. She already begins the fulfillment in all those expressive movements which make her attractive through the irresistible demonstration of her feminine bodilihood as concealing revelation of the "great mystery." Is this explanation of the sexual attractiveness of the woman too "romantic"? Through his art the poet evokes images which were motivated by human experience. If his work of art is to address itself to us, the non-artists, then we must be able to recognize these experiences in his image. The poet shows us basic existential structures – without unfolding them as the philosopher does – through which the activities of our everyday life appear in their *genuine* meaning.[92] He does so in his own way, and particularly through his *poetizing* which not only means a re-creation of the human drama, but which makes this drama at the same time transparent with respect to the values which decide the meaning of this drama. These values are shaped in mythology where they become "persons" who live in their own "mythical" world. Thus he reveals and celebrates the archetype of the woman and makes the concrete beloved one transparent with respect to this archetype. In the everyday sexual encounter this does not take place – although every-

90. G. Marcel, *Du refus à l'invocation* (Paris: Gallimard, 1940), p. 123.
91. Cf. Binswanger, *op. cit.*
92. Cf. F.J.J. Buytendijk, *De psychologie van de roman* (Utrecht: Het Spectrum, 1951).

one more or less becomes a poet there. In the everyday sexual encounter one does not experience more than an ecstasy over the feminine figure in which what is hypothesized by the poet addresses itself to us directly. If we keep this in mind, it is easy to understand that the poetic outpourings have great value for our insight into this encounter – greater value than the testimony of the "normal" lover who only knows how to say that his beloved is *beautiful*. "She is made like a pretty girl; what dost thou wish me to say? Must I name for thee every part of her body, adding that it is beautiful?"[93] Thus we do not attribute a "metaphysical value" to the poems of the poet; we interpret them with respect to their *psychological* meaning, but recognize in so doing that poetic description as far as love-matters are concerned often speaks a clearer language than any prosaic analysis. For it is just not possible to explain a sentiment in rational terms. The language of poetry which represents in images tries to make explicit in its way what addresses itself to us in our everyday experience. Here we may speak of "the great mystery" of the woman without burdening our scientific conscience.

The very harmonic relaxation of her body makes a woman more mysterious and more attractive. And to a greater extent, therefore, when she is asleep. And this point brings us to the explanation of the thesis that her more direct way of being-bodily means a higher degree of appearing-interiority.

That a woman is more beautiful and desirable in her sleep is an "old truth" which is established time and again: "the sleep of this woman confused me – because like all extremely simple things it was impossible for it not to conceal a secret."[94] For Marcel Proust, sleep is the necessary condition under which he is first able to love Albertine genuinely.[95] Why? Stretched out at full length upon his bed, in an attitude so natural that no artist could have painted it, she had become a *plant*. She was animated now only by the unconscious life of plants and trees. She had called back into herself everything of her that had originated from her; "she had taken refuge, and was enclosed and reabsorbed in her body." That is why it seemed to Marcel "that there was, condensed in it, the whole person, the whole life of the charming captive, outstretched there before my eyes." In the sexual encounter the question is about the other, and this other person, Albertine, seems to be completely present here; nothing escapes him, the totality of her inner life, no longer filled with eventual secret intentions, expresses itself in the "vegetableness" of her sleep. That is why he loves her. "Her ego did not escape at each moment, as it did when we were talking, through the channels of her unacknowledged thoughts and of her gaze." There is no longer any hidden and secret interiority, but there is an all the more mysterious interiority. For the atonia of sleep is "the expression of an existence without intentions, but because of this also the expression of something positive, namely of a being-with-oneself, of an interiorization."[96] It is a revelation of the great mystery of beginning and end, of a carefree life in the

93. Louys, *Aphrodite*, p. 82. (English: pp. 58-59).
94. Flaiano, *The Short Cut*, p. 30. Cf. *De vrouw*, pp. 232ff.
95. Cf. M. Proust, *La prisonnière*, vol. I, pp. 84-85.
96. Buytendijk, *De vrouw*, p. 232.

security of the "home," of the luxury and abundance of a life which re-creates itself time and again. The concrete beloved one certainly does not show us this. But if we encounter her as a woman in the pregnant sense of the term, then all of this seems to be included in her expressive gesture, in the fulfillment of her body. However, and this is important, it is contained in it only as a *promise*. Sexual desire does not require that this promise be fulfilled; it merely asks that the woman promises all of this *because of me.*

Each form we perceive has the characteristic of mysterious interiority, in that it does not have a cause, and represents a value in itself which nevertheless seems to be hidden behind it. We say, seems to be hidden, because there can be no question about a really hidden entity. When we encounter the woman as a mystery, we encounter her as representing "an independent and meaningful being, which exists in itself as completed and closed and possesses an *immanent* meaning which it cannot abandon."[97] The woman addresses herself to us with a value which we cannot understand from the situation, nor from our everyday world. This value *is* not hidden "in" her, but *addresses itself to us* as a mysterious interiority. For she becomes present in her figure without the value itself coinciding with the body. That is why in the eyes of the spectator an interiority forms itself and seems to thrust itself to the fore until it reaches the surface of the body, and then time and again recedes behind this surface.

The "hidden interiority" was mentioned earlier. Now is the time to compare it with the "mysterious interiority." For these two concepts are not congruous. Plans and intentions remain hidden only insofar as the person does not materialize them. Hidden interiority is the hypostatization of a person's *intransparency*. The manner in which he is present to me is ambiguous. I understand his intentions from his activities which in turn I understand from the situation. However it can be that he hides his "true intentions," or that I misunderstand him. But the hidden interiority can, albeit not always in fact, nonetheless in principle be fathomed. It correlates with "masculine," goal-oriented activities, thus with a mode of existence which is oriented toward the world. What is hidden can also be understood as a *secret* intention, which although oriented towards the world, is now kept back and concealed. The mysterious interiority, on the other hand, is an "inexpressible" interiority, one which cannot be materialized in goal-oriented activities and therefore is inscrutable. Taken in itself each expression witnesses such an inscrutable interiority. It can be understood insofar as it is not taken in its immanent value, but merely as an epi-phenomenon of the activity which gives meaning to the becoming-present of the person. Any expression which frees itself of its epi-phenomenality, becomes at once inscrutable, enigmatic, referring to a mysterious interiority which expresses itself in it without really revealing itself.

Insofar as the body is given as an incorporation of an immanent value which is not oriented toward the world in order to become involved in it, it already appears for that very reason as a mysterious medium with a pre-given expressive essence. Psychologically seen it is a *feminine* body insofar as it incorporates this expressive

97. Buytendijk, *De vrouw*, p. 229. Cf. the entire chapter on "mysterious interiority."

essence. But in addition to being the body of a woman it is also a *human* body, and as such equally is as instrumentally "useful" as that of the man. However, through her appearance, her figure, she seems to be destined not to materialize the masculine, instrumental form of existence. She fulfills the predelineated meaning of her body insofar as she does not become present *as* a person who is oriented *toward* the world and receives his orientation from the world, but precisely as a person who sojourns within herself. In so doing she accepts the meaning of this body for the other and thus becomes woman in the pregnant sense of the term. This happens in the erotic gesture which as we saw is "useless," that is oriented neither toward nor by the world. This is why through this gesture there manifests itself for the man a mysterious interiority. It is now no longer her body which appeals to him, but her "soul" as inner landscape. Through the erotic gesture the woman invites him to enter this landscape as the born discoverer, traveller, and conqueror. This explains the myth concerning the man as "the conqueror of the woman's heart." However she does not invite him to enter in a "masculine" manner, that is as *curious*. The meaning of the inner landscape becomes manifest to him not insofar as he is a curious investigator, but only as a lover; otherwise he will end up the victim, dupe, and being disappointed – *tristis*. The woman invites him as a *man*, but he must enter as a *lover*, who does not search for the usefulness but for the value of the beloved.

The woman *materializes* her body through the erotic gesture as a *tender* gesture. Yet the erotic gesture taken in itself does not suffice to sexualize her body; therefore we must still look for a final specification. For even the smile and posture of the Mother of God are erotic gestures in the sense we have given to the term. She becomes present in these gestures as a Woman, who completely materializes what is feminine and constitutes herself as a person in harmony with the predelineated meaning of her body. If She does not address Herself to me sexually, then this is because Her gesture does not address my "earthly" existence. Pointing to the Child. She calls me to *another* form of existence; She points to a world which is "above" this world.

The erotic gesture becomes sexual as soon as it addresses itself to *me*, this concrete "earthly" person. That is to say as soon as in the erotic gesture the other person, the woman, offers her tenderness to me personally, invites me to share her existence with her as a person – and thus gives herself in the sense of being willing to receive me, willing to receive me *bodily*. In the gesture which addresses itself to me sexually, the woman already begins to give herself, incarnate herself, and asks me to do the same with her and to complete this in a dual incarnation. Thus we can understand why the man is called the "seducer" and why the woman nonetheless is "blamed" for the seduction. In her gesture she invites him to enter (guilt), but he must *do* it, or as one says, he must take it up (seduction).

If she loves her man this is because of his power and pride, his decisiveness, and his *personality*. But she does not expect him to enter her world in the execution of these qualities. In the sexual encounter she laughs over his masculinity, calls him "a great child" and a "silly boy"; she expects his tenderness to be in union with hers. We have seen that the kiss, the caress, the erection are *expressions*, manifestations of tenderness which find their fulfillment in the union of the bodies and the unification of their "souls" which playfully they intend therein. Tenderness is the realm of the woman, the basic characteristic of her world and personality. On the basis of what has preceded we may call this realm without fear of confusion the world of *motherliness*, and we can understand with Buytendijk "that motherliness as a respectful and careful being together in disinterested giving, as a warming, nourishing, cherishing, and caressing, is the power which everywhere and always – in men, in nature, and in culture – *evokes* what is hidden, tender, subtle, fragile, and germinating."[98] Thus tenderness is *more* than a desire to touch and be touched. There is also a curious desire to touch that just wants to know "how it is." Tenderness, if it is not disinterested, is certainly without intention, without pre-established aim; is not oriented toward knowledge, benefit, lust, or result; tenderness "just is," because of you: "nothing exists, nothing counts, except those eyes, that forehead."[99] Tenderness is a quality of the expression which is supported by surrender. This is the meaning of the woman's body in the sexual encounter, with the specification that the woman gives *herself* to the other for the sake of the other. In this light there is no longer any mysteriousness about the greater expressivity of her body, her so-called passivity, nor the suppleness and naturalness with which she knows how to give herself. Her body fascinates: it is tenderness which has become incarnated for my sake. Her body *is* her tenderness, is she herself. Seeing this body a man cannot do anything except try to soften his hardness – and his body which is "wiser" than he himself has then already started this.

X

The sexual encounter finds its "source of energy" in the polarity of the sexes, that is, in the polarity of the masculine and feminine worlds. Saying this we return to a starting point which was rejected earlier, and feel that van den Berg's statement that one must have gone shopping with his wife on Broadway before he can write a sexuology is justified.[100] Before any sexual encounter in the narrower sense, the polarity of what is masculine and feminine is already given as the polarity of *the hard world of work and the tender world of "care" and "sojourning."* But it is equally certain that our sexual desire cannot be put on a par with the revelation of the specificity of a world *other* than one's own. This discovery is a *condition*. Condition is also the orientation toward the other as a partner. Sexuality manifests itself then as the attractiveness of the other, and simultaneously as the lived incarnation in one's own body. Here we thus maintain our position that sexuality in the narrower sense does not *exclusively* become manifest in the landscape and, thus, can be understood only through an analysis of the sexual encounter as reciprocal incarnation.

98. Buytendijk, *De vrouw*, p. 334.
99. Géraldy, *Toi et moi*, p. 58.
100. J.H. van den Berg, "Het gesprek," in *Persoon en wereld* (Utrecht: Bijleveld, 1963), pp. 136-154.

This encounter tends toward an abolishment of the polarity, namely by fulfilling the feminine world and transforming the masculine. And so we find back again Plato's idea that Eros is nostalgia for the "original" unity of the sexes, for the one and undivided existence, which, however because of the absence of inner tension, would no longer be a real existence. By looking at it in this way we accept the sexual differentiation as an original possibility of encounter which finds its "natural" fulfillment in the encounter of the sexes. Love *can* be love for someone of the same sex, and *can* find its fulfillment in a homosexual encounter. If such a relationship appears to us as unnatural and even as unacceptable, it is not because in such a case one could not speak of real love. A man *can* love homosexually, *but in so doing he can never fulfill the pre-given meaning of the differential bodilihood*. The biological fact of the differentiation of the sexes receives a human meaning here. Man and woman are oriented toward one another. Psychologically we can*not* justify this by referring to their anatomic construction. Psychologically we account for this by referring to the polarity of their worlds and the polarity of the masculine and feminine modes of existence given with it, modes of existence which supplement one another but can never abolish each other. This polarity is pre-given in the bodily figure, not as a biological fact, but as meaningful figure, namely as the medium through which the human person already "by nature" becomes present as man and woman.

It is the task of philosophical anthropology, as the doctrine of man's essence, to show that sexual differentiation taken as a polarity of the masculine and feminine worlds belongs to the very essence of our being human. Psychologically this polarity is irreducible. Psychology accepts these data. It is not its task to show that the body must be understood as the medium of one's existence, no more than it is its task to explain the sexual differentiation. It takes these – as we have seen in this essay – as a point of departure in order to explain the concrete sexual encounter as a psychological phenomenon.

In the encounter the other is not given to us as a body, but *through* his forms of behavior. In each concrete form of behavior we find a unity of action and expression. Usually we understand our behavior more as an action, as goal-oriented behavior, and the other becomes present to us through this behavior as a person "who is after something": who intends something. The total image is then understood from this intention. We understand what this intention is from his meaningful concern with things in the situation. But sometimes we do not understand his forms of behavior as actions: namely, because certain forms and patterns of behavior do not seem to flow forth from an intention. We then speak of expressions or expressive movements and then conceive of these as expressions of something which goes on in his inner life,[101] of an *emotion*. But although it is true that this expression is not a goal-oriented form of behavior, it is not without any relation to the situation. It expresses the manner in which the other experiences his involvement in the situation. Thus in the expression we discover the experience of the other and, therefore,

101. Cf. F.J.J. Buytendijk and H. Plessner, "Die Deutung des mimischen Ausdrucks," in *Philosophischer Anzeiger*, 1925-1926, p. 90.

in a certain sense *himself*, just as he is in his so-being. The expression is not directed toward something, nor through something; and so it is not an *action*. It is not even an "unconscious" action; the person neither consciously nor unconsciously *wishes* to make something manifest through his expression. His expressive movements are spontaneous. A man exists through his body and this shows the signs of the manner in which he is in the situation. If the expression is conceived of as a *result* of an "inner state," then it is considered an *action*. One can do so; but one can also conversely conceive of an action as an expression, as Merleau-Ponty does when he says that "the body at each moment expresses the existence."[102] If we exist acting, then each action is the expression of this existence. Action and expressive movement therefore are not "objectively" different forms of behavior, but *manners in which we understand the behavior of the other*. A form of behavior appears as an *action* when it is understood in the encounter as a *function* oriented toward and through the situation. It appears as *expression* if it is understood as the expression of a meaning which is contained in the movement-image itself, a meaning which is the expression of the manner in which the person experiences the situation. "Behind" this expression we expect to find the one who expresses himself so: the person himself who *is* just as he expresses himself. In the action he alienates himself from himself, in the expression he is himself. Sexual desire asks for the person and thus asks for expression, not for his activities. In the sexual encounter all forms of behavior function as expressions, and this means that the body does not function in its instrumentality but rather in its expressivity, and this is so because in the sexual encounter the body is understood in that way and must be so understood. In other words this is so because of the fact that the intentionality which underlies the sexual encounter and forms it, is not a goal-oriented intentionality, not an action-intentionality, but an emotional and affective one, not an intentionality focussing on one point but a diffuse one; and because of the fact that the sexual encounter itself is not an explanation but a being-together, in which both partners experience the ambiguity of one's own and the other's body. They *incarnate* themselves not only in the eyes of the other but also in their *own* experience. Experiencing and acting man lives in a way in which he passes over himself in his orientation toward the world where the gravitational center of his action and experience is to be found. In his encounter with the other, too, he passes over his own body in that it functions merely as a *means*. The incarnation is usually experienced in the other: we see him becoming present through his body. In the sexual encounter we feel ourselves, too, becoming present in our own body. My own incarnation is motivated by the specific manner in which the other incarnates himself, namely as a person who in the erotic gesture invites me to enter his intimate world. That is why sexual desire is a desire that already originally includes the other, although the possibility remains that the experience of one's own incarnation fascinates him so much that he forgets the other. Sexual desire emerges as a *circular process*.[103] When van den

102. Merleau-Ponty, *Phenomenology of Perception*, pp. 165-166.

103. Cf. B.J. Kouwer and J. Linschoten, *Inleiding tot de psychologie* (Assen: Van Gorcum, 1951), Chapter 5.

Berg following Buytendijk observes that there is no desire without seduction[104] and thus places the center of gravity of this desire in the seducing object, one can say conversely that there is no seduction without desire. Desire and seduction belong together, motivate one another reciprocally, and are so interwoven that it is difficult to distinguish between them. Person and landscape find their unity in the situation. That is why we can analyze desire best as a situation such as it manifests itself in the landscape of seduction. But the psychologist sees how the person projects the landscape and in this way situates himself. That is why ultimately he attributes the desire to the person as the bearer and origin of the situation. The sexual incarnation of the other is given to us in his expressions insofar as they can no longer be understood from his actions, but merely as the expression of the inner form's becoming present without any intentions. For the woman this means the fulfillment of the pre-given meaning of her body, which is always understood as a "properly" expressive medium. For the man this means a transformation of the pre-given meaning of his body which is understood as an instrumental medium. For both this means a falling away of the world insofar as it appears as the field of all goal-oriented forms of behavior. In the sexual encounter the world becomes now a common emotional field in which everything changes its meaning. This is how the "place" becomes constituted in which we can encounter one another without misunderstanding. The body of the other as well as my own body command more and more of my attention. I coincide as a person more and more with my own body and thus lose the distance from myself and from the world. Finally, in the orgasm, my body is my consciousness; a "consciousness made flesh" which "lost" it environing world; it experiences merely the other becoming present and its own becoming present toward the other. The world has not disappeared, has not dissolved, but it has become *condensed*:

> The world had disappeared at their boundaries,
> They were merely one another's partners
> Dreamless taking part in one another's members....
> ....
> The sweet birds fall silent. The world
> Withdraws up to the skin of the lovers.
> The last fire of the world forges together
> In the unspeakableness of their embrace.
> The universe becomes empty around their couch,
> Devoured by their nameless kisses,
> Devoured by their unity, breast to breast.[105]

Thus not only the distance from myself, but also the distance from the other becomes minimized. The limit of this loss of distance in regard to myself is the lived unification with the other. Even the appearance of the other is still too much; the consciousness-made-flesh can maintain itself and complete itself only by receiving

104. van den Berg, *Menselijk lichaam*, p. 406. Buytendijk, *Het spel*, p. 81.
105. Aafjes, *op. cit.*, pp. 27-28.

the other, by blending together with him in a dual incarnation. This is possible only through the unification of the two bodies as partial-bodies, by becoming present together as part-takers. My intimate existence and your intimate existence melt into one intimacy in which our bodies have no other function but to express this intimacy.

But we "know" very well here that our bodies are *two* and that we, however must try to materialize ourselves as *Wirheit* (We-ness), though our bodies will always *remain two*. That which makes our being-together possible, hinders our unification. That is why the unification of the persons is a "*played*" single-personality.

From this the expression "love-play" receives its genuine meaning. In the fond word, in the kiss and caress, and in the embrace we move, playing, toward one another and let ourselves become present for one another and together, as if this could lead to a genuine unification of two persons in one existence. But then the love-play in the narrower sense is no longer "preparation" for sexual intercourse, but conversely, intercourse is included in the play of love, this encounter in which we meet one another not as adversaries and not even as opponents, but in which we as play-mates are together. That is why the "act" of intercourse does not have the structure of an *action*. Taken as a form of behavior we *can* regard intercourse as an action; but taken in itself it demands to be conceived of as a dual expression. The body never becomes completely an expressive medium; in the sexual incarnation it *strives for* that. Since it never can get that far, this limit-meaning of the body is played in the love-play. This play was described and analyzed here in its basic structure just as it comes to the fore through the meaning of the body in the sexual encounter. This does not mean that sexual relations cannot be different than we have described them. On the contrary, an infinite variation of constitution is here concretely possible. On a few occasions we have indicated some of these possibilities. But any variation is a variation within the leeway which our body, this *provisional sketch of our existence*, leaves open for us.

The body as ambiguous medium appears here in a threefold manner: (1) It appears first through the loss of its instrumental part toward which it is oriented in our everyday existence. The body of the man becomes "silly" here, at least for the ironic spectator who sees how difficult this transformation often is for a man. The body of the woman on the other hand becomes precisely "what it genuinely is": an expression of her feminine personality; and this fulfillment has enchanted the poets for many centuries – and not just the poets. (2) Furthermore the body is ambiguous as the "place" where in the sexual encounter the outer – taken in the literal sense – and the inner flow together – albeit never completely. There is always the possibility, to point to an extreme, that the expression is no longer an expression but an *activity*, and tenderness is no longer genuine and spontaneous, but simulated. (3) Finally the body is ambiguous where, as a fulfilled expression of one's personal existence, it is merely the medium for a dual incarnation, but simultaneously the barrier which prevents a complete union. However much our personal existence taken in its basic structure is oriented toward the other, because of our body we exist as individual persons and never as "Wirheit."

That is why the sexual encounter remains under the sign of what Valéry has hinted at: By nature we are condemned to live in the imaginary, in that which can never be completed. But in this "condemnation" there is at the same time the possibility of the sexual encounter as the played fulfillment of our personal existence in a participation in the other person. This implies also that the sexual encounter is: *a song of desire*, always new and always fascinating. For the desire is never exhausted and the promise is never fulfilled. The sexual encounter is a game of desire and promise which do not wish to be more than that. In this game what is *mostly my own* becomes *yours* through my surrender; it becomes a medium in which beneath your look and caressing hand I become present to receive you.

> sed sic sic sine fine feriati
> et tecum iaceamus osculantes.
> hic nullus labor est ruborque nullus:
> hoc iuvit, iuvat et diu iuvabit;
> hoc non deficit incipitque semper.[106]

106. Petronius Arbiter, († 66), quoted by H. Waddell, *Medieval Latin Lyrics*, Penguin Classics, L. 29, 1952, p. 24. Freely translated this becomes:
"But let us celebrate this, just in this way, without an end,
Together kissing stretch our bodies.
Here there is no labor and no shame:
Here everything is, was, and ever will be good
Here nothing breaks off, and everything will always begin anew."

Experienced Freedom and Moral Freedom in the Child's Consciousness *

F.J.J. BUYTENDIJK

No other word has so much power to divert the mind of man from the fatigues of toil, from cares, from the hazards of emotional involvements, from all selfishness and meanness, as the word freedom. The magic power of this word is so great that the burning desire for freedom is not merely an appeal to a well-determined concept, but it opens a door to another climate. In the name of freedom the most sublime sacrifices and the most revolting injustices have been committed. No one knows exactly what freedom is, but all consider it as the sovereign good. Humanity's history and each particular man's history are exclusively determined by their relation to freedom.

When Bergson declares[1]: "Freedom is a fact, and among the facts we know, there is none more clear," the word "clear" does not here mean the clarity of an intellectual thought, but that of a cloudless sky. It is the clarity of pure spirituality, which does not let itself be grasped nor fixed by any concept. If freedom is the supreme good, if it is man's dignity and his ultimate goal, we may easily understand that any educator must of necessity set for himself the problem of learning how the child may be made to participate in this supreme good.

The aims and ideals of education have changed in vain – the tendency to give to man his most noble form remains invariable. That means our aim is to give him a free existence. However, the ethic and the pathetic of education are threatened with losing their animating force if we ignore the conditions which must be fulfilled in order to bring about true freedom, moral freedom.

In order to know the general and basic nature of these conditions, we must examine the development of the consciousness of freedom in the child. That is why I take the liberty of drawing attention to the relation of experienced freedom and moral freedom. To this end, we need to answer coherently the following questions.

* This article was originally published in French, "La liberté vécue et la liberté morale dans la conscience enfantine," in *La Revue Philosophique*, January-March, 1951. The translation was made by Alic Weymouth. The English translation appeared in *Educational Theory*, 3, no. 1 (January 1953), pp. 1-13. Reprinted by permission of the Editor-in-Chief.

1. Henri Bergson, *Essai sur les données immédiates de la conscience* (Paris: Presses Universitaires de France, 1944), p. 166.

In what situations does the consciousness of freedom appear for the first time in the child? How is this primary experienced freedom anchored and rooted in the unconsciousness life? Can experienced freedom be changed by itself into moral freedom? Is the basis of the latter also in the original relations of the child with his environment? How can education reinforce moral freedom so that it becomes a life experience of the child, integrated to the totality of his personal development; that is to say, by what means will moral freedom be vitalized or life entirely humanized? I may be permitted to add here an introductory remark. Morality is often nothing other than the frame or prison of the autonomous life.

Only too often we choose to display a wild and troubled life in this embellished and gilded frame; morality adds luster to the gallery of our honorable contemporaries; and we even find in the historical museum of honest humanity many a portrait of a celebrated man furnished with a moral framework.

Even more often, through deficiencies in education, we experience morality as a hindrance of freedom of movement. That is why man poisons his own existence by a melancholy and sentimental nostalgia for an imaginary, dreamed-of freedom, that of the wild animal, of the bird, of all nature. This caged existence is ever ready to break its bonds and escape toward libertinage, apparent freedom which is in reality the enslavement of the passions.

Madame Montessori deserves great credit for having reformulated education in such a way that henceforth morality may become for the child an experienced and assumed freedom and no longer a prison or a framework for display purposes. I need not re-tell how Mme. Montessori attained this goal. The education in moral freedom which she set forth in her books, which is in daily use in her schools, is based on a vast and profound empirical experience of the child, as it is revealed in the different phases of his development.

As a theoretical psychologist, I have felt that I should limit my task. It is not within my competence to pass judgment on questions relating to pedagogy. Nor would I wish to venture into the vast field of empirical child philosophy. I wish only to try to clarify the relationship between experienced freedom and moral freedom, in the light of the questions posed above, starting from the essential character of the child's existence. Thus I hope to be able to contribute, in some small measure, to a clarification of the theoretical basis of education in true freedom.

But first we must state how, in our sense, the problem is related to the sciences and to philosophy. The positive sciences as well as philosophy often present a serious danger for education, a danger which is far from imaginary. The proof is seen in the way the fundamental principles of education have been fought and contested. We have had, of course, in this discussion often to depend on the authority of results in biology and physiology; but it has often not been well understood that the child – although he differs very much from the adult – must be considered, first of all, as a human being. That means that he is not merely an example of a natural species, but the representative of an historical idea.[2]

In any case, the child is specifically human from the moment when he first turns

2. Maurice Merleau-Ponty, *Phénoménologie de la perception* (Paris: Gallimard, 1945); English

his gaze toward people and objects and when he expresses the blossoming of his human nature in his first smile. Already, in that hesitant response, the child is building for himself the beginning of his being, and at that moment he surpasses the limits of unconscious life and the laws of nature.[3] Only the teacher who takes account of the human nature of the child, expressing itself in his existence from the beginning, will be protected against the danger of biologism and naturalism. The positive sciences will never be able to convince such a teacher that the human being, in his earliest years, is only an organism which develops and reacts to the stimulations of the environment.

From the moment we discover human reality in the child, we also meet the irreducible phenomenon of limited freedom. From the moment when consciousness is aroused, there is a choice of world, a projection of world. "To be born, is to be born *of* the world and *to* the world at the same time...." "Thus there is never determinism," but on the other hand there is "never absolute choice." ... "The idea of situation excludes absolute freedom at the beginning of our operations." ... "We choose our world and the world chooses us."[4]

Modern education should take these thoughts on the essence of human reality as directives, in order to be able to approach, at a more favorable time, the old problem of free education, of freedom. Now the result of this way of doing is threatened not only by the inadmissible parallel between the development of the child and animal life, but also by the danger of an unnecessary confusion of pedagogy and metaphysical problems. Although we should recognize that philosophy is capable of discovering the "essentia abstracta" of man, the "essentia concreta" never precedes existence. Man is always what he does and what he has done in his intentional relations with the world.

Philosophy can of course acquaint us with human possibilities, but it can never reveal the reality of the existence of a concrete man. And it is precisely this reality of existence which we must know if we are called upon to aid the development of a child to his true dignity as a human being. Thus we may see why the problem of education directed toward freedom cannot be clarified by a metaphysical discussion of arbitrary freedom or determinism or indeterminism. All philosophic speculation is sterile in an investigation of the relation of experienced freedom to moral freedom. We must rather be oriented toward the concreteness of the child's existence.

Concerning this existence we may state that it is formed by autonomous activity, the initiative of the mind. The child, like all mankind – of whatever age or condition – is essentially characterized by the fact that, through the manifestation of his freedom, he is given his nature as well as his grasp of the world, his ethical principle, and consequently his system of values.

At the same time that the child is forming his own world, he is giving a direction

---

Translation by Colin Smith: *Phenomenology of Perception* (London: Routledge and Kegan Paul, 1962).

3. F.J.J. Buytendijk, "Das Lachen des Kindes," *Psyche* II, p. 57.
4. Merleau-Ponty, *op. cit.*, pp. 517-518 (453-454).

to his physical being and lending it a power which the body does not naturally possess. The choice which the child of necessity makes in the different situations which his environment presents to him must be confirmed, corrected, or revoked at every instant of his life.

Thus D. de Petter[5] is right when he says that the essentially educative act is that of the subject himself, by which he himself is formed, in his vital and autonomous activity, by all which he has recognized and appreciated as having a value in itself. These considerations show us clearly that education differs, because of the necessity for the fulfillment of moral freedom, from any form of training. Neither the positive sciences nor metaphysical speculations have any decisive influence on our research.

Must we then conclude that psychology is the science which can guide our search toward a clear understanding of the meaning of freedom? In order to reply to this question, we must start from the distinction between the classic psychology of the textbooks and treatises, and the knowledge of man initiated by immediate participation and communication, in existence in the different phases of life and in diverse situations.

Classic psychology is based on the discovery and description of psychic phenomena, of behaviors, and their statistical regularity. The attempt is made to explain these phenomena by hypotheses related to a psychic mechanism of the inferred interiority. As important as the results of these studies may be, they do not succeed in giving an explanation of the development of will and self-consciousness, especially if we wish to limit ourselves to empirical facts. I would do better perhaps to illustrate this by an example.

A well-known psychologist, P. Guillaume,[6] declares: "The child becomes conscious of himself only at that moment when, through imitation, he sees himself in the person of another." But how can we know whether this explanation is valid, that is to say, whether it is truly the revelation of intelligible relations between the child, his self-consciousness, and the world?

The truth would appear only if we knew what self-consciousness is, in the proper sense of the word. But self-consciousness is not a thing to be proved, and one cannot even define it. We can understand only the act of direction toward one's self, because we are capable of doing it in the same manner. In addition, if we wish to take full responsibility for our concepts, we must ask ourselves: what, in reality, is "to see one's self?" And what is the real meaning of the act of "seeing one's self in the person of another," and in what way is imitation a "seeing one's self," and in what situation? In brief, modern psychology differs from classic psychology in that it takes account of the precise meaning of concepts.

In any case, the application of this method, called phenomenological, is absolutely essential for a profound understanding of freedom, both experienced and moral. Because of the discursiveness of our knowledge of behavior, we are obliged

5. D.M. De Petter, in *Studia Catholica*, 51(1948).
6. P. Guillaume, *L'imitation chez l'enfant* (Paris: Alcan, 1925).

to try to penetrate, by means of an eidetic intuition, into the essence of the basic acts through which the child forms his existence.

In the field of education, Mme. Montessori in particular glimpsed, through all the empirical data, the essential characteristics of the child, which make the whole empirical experience intelligible. The fundamental characteristics which, according to Mme. Montessori, throw light on the whole behavior of the child, are spontaneous initiative and that aspect of existence which she has designated by the insightful concept of "the absorbent mind."[7]

We misunderstand the meaning of this concept if we conceive the child's mind as a sort of sponge, which, by virtue of a pre-existent structure, should absorb everything which enters it through the apertures of the sense organs. Such a concept of consciousness would conform to that of classic psychology, so felicitously described by Jean-Paul Sartre[8]; "... We have all believed that the Spider-Mind drew things into its net, covered them with a white slime, and slowly digested them to its own substance." The concepts of Mme. Montessori are radically removed from this "alimentary philosophy."

We believe we understand her correctly when we say that her whole method of education is based on the *human* reality of the child, namely, freedom of initiative. The absorbent mind is nothing without the originality of this initiative.

"To be, is to burst into the world; it is to come forth from a nothingness of world and consciousness and suddenly burst-consciousness-into-the-world."[9]

The human being of the child is fulfilled only through his existence, his being *in* the world and *for* the world, through his physical being. In this existence, the consciousness of the child becomes at the same time consciousness of the world and consciousness of self, that is to say, consciousness of his initiative, of his autonomy and of his independence.

Doubtless, the child lacks, at the beginning of his life, any clear image of self and consequently of his freedom, but he discovers himself to be free even in his random activities, although he has only a quite nebulous and diffuse sensation of this freedom. This consciousness is the first consciousness of experienced freedom. The soil in which this freedom can take root is offered to the child by the body and it is nourished by the animal life in which he participates.

That is why this first consciousness of freedom is so strong in the healthy child. In him, vivacity and a spontaneous bent toward movement are clearly manifested. To be more than a feeling of force and health, to become an experienced freedom, this vital spontaneity, at this stage of the child's development, must be considered as a possible refusal. It must first of all mean a resistance to being governed, then later an upsetting of the order of things, and finally it must mean a revolt against the order. Let us try more precisely to define this first way in which experienced freedom is presented to us.

7. Maria Montessori, *The Absorbent Mind* (Madras: Ganesh, 1949).
8. Jean-Paul Sartre, *Situations*, 6 vols. (Paris: Gallimard, 1947-1965) vol. I, p. 31.
9. René Le Senne, *Traité de morale général* (Paris: Presses Universitaires de France, 1947), p. 27.

In the first place, we may state that this first phase of experienced freedom is not the consciousness of true freedom, that is to say of that freedom "indissolubly, essentially attached to the mind," and which "is real only when it defends itself and conquers itself." There can be no question of this in the first phase.

In the second place, we may state that primary experienced freedom depends on emotional dynamism. Le Senne has thrown light very well on the difference between freedom of initiative and freedom of power. This latter is realized only in proportion as reason and, generally, value, confer it. "In this sense, Erostratus, who set fire to the temple of Ephesus, was not free; he was the slave of his passions and of his errors. – He was free in the sense of freedom of initiative – he was free in the sense that he would have been able not to set fire to the temple."[10]

It nevertheless seems to me that Erostratus was not entirely determined by his passions, but that he chose rebellion and revolt against a reasonable moral order and that he wished to set them in the context of a world projected in its totality as refusal. This projecting of a world of refusal, in which the person becomes involved and to which he submits, means the choice of primitive experienced freedom. Thus we define it more precisely as the *freedom of caprice and of revolt*.

There is in every man, and certainly in every child, an almost unquenchable need for the experiencing of this irrational freedom – a freedom even senseless and absurd. We recall that Dostoyevsky, in his *Memoirs of the Underground*, showed us the tragedy of this apparent freedom. The consciousness of initiative may easily evolve into consciousness of caprice, as a means of attaining the experience of a feeling which is commonly called freedom.

There is no one who is not inclined, from time to time, to assume the role of a child in revolt, in order to do "what he wants to do." Thus we are not at all surprised that the child keeps this tendency so long, if an appropriate environment and a judicious education do not offer him the means of attaining a higher stage of freedom. However, it is common knowledge that *all* children do not use revolt as the means of acquiring the feeling of the experience of freedom.

There is general agreement in recognizing that it is especially small boys who revolt, that they are more apt to follow their caprices and that they more often attack the order and value of things. It is evident that this difference between the sexes, which is manifested even at an early age, rests on a difference of innate disposition and of education during the first years of life.

Elsewhere I have tried to show that the innate difference is manifested in the fundamental form of the dynamism. Dynamism displays a predominant tendency either toward expansion or adaptation. For this reason, the very first contacts of each sex with the world differ.

The expansive movement causes us to encounter the world as an obstacle; the adaptive movement, on the contrary, causes us to discover a world of qualities, of form, of values. The meeting of obstacles arouses, by reaction, a reinforced expansive dynamism. Thus the child, physiologically stimulated, develops a greater

10. René Le Senne, *Introduction à la philosophie* (Paris: Presses Universitaires de France, 1939), p. 49.

muscular tension, and is thereby more predisposed toward a more aggressive behavior.

This aggressiveness is not itself an innate and physiologically determined quality. It is not an inevitable outcome. Does not the teacher himself impose on the "real" little boy, strong and energetic, different demands than on the little girl; therefore he stimulates aggressive conduct, or at least he does not modify it as much as he could.

Whereas the little girl is especially praised when her movements and occupations help her conform to the traditional image, fitting for her age, we demand, on the other hand, that the "real" boy *do* something, no matter what, so long as he acts. We prefer to see the boy, in his approach to the world, as a builder, as a conqueror, as a dominator, a little *homo faber* (*homunculus faber*). In my opinion, this opposition of the sexes, artificially cultivated from early childhood, is one of the principal faults of education. Fortunately, the last generation has made appreciable progress; and especially in the Montessori schools, each child, boy or girl, finds the appropriate environment for the development of expansive tasks and an adaptive relationship with things. However, at the beginning of life, all activity is determined negatively rather than positively. This may be explained by the fact that the world is not offered to the child primarily as a structure full of meaning, but as an obstacle and an opposition with relation to his expansive spontaneity.

This expansivity is in itself blind to all value. This means that it has, chiefly, a negative relation toward all reality and that this relationship immediately becomes self-conscious in the form of revolt, as soon as anything energetically resists this expansivity. The *thing* may be an object which bars the way in the true sense of the word, but it may also consist of an order, a command, an imperious look, existing *as such* for the child. But there exists nothing for man, and consequently for the child, which may not be projected, in its structure of meanings, by an intentional act, or by an instance of perception, of refusal, or of acceptance.

From the moment we understand clearly this fundamental characteristic of human existence, the relation and the difference between experienced freedom, which is primarily nothing other than freedom of caprice and of revolt, and moral freedom, are already somewhat apparent to us. Thus we may note that experienced freedom cannot be expanded except through opposition to a demand emanating from the world – that is to say, from things and people – which is understood as such, but which is refused. Caprice and revolt are possible only with respect to a value which seems to announce itself to be irresistible. Thus we have arrived at a first conclusion.

The small child who breaks an object, tears it up, destroys it, who resists, who does "what he wants," is not, animal-like, following a non-ordered and non-directed impulsiveness. The child gives to its impulsiveness, also aroused in him by physical stimulation, the meaning of a revolt. Therefore his acts become really a revolt with respect to the positive values of being, discovered and denied as such in refusal. The child is, while still very young, a true man, and for this reason he is not a representative of a nature obedient to blind laws, but always that of a normative and consequently moral consciousness.

The mother, in her intuitive wisdom, is thus to a certain extent right when she says that the baby is "naughty" when he *gleefully* throws his plate on the floor, watching it fall with evident pleasure. However, the child who takes the initiative which first wells up in him without interior order, from his impassioned vitality, is awakened to a first human reality, to which he has a right. In this awakening, he encounters things in their apparent autonomy, their mass, their weight, their inertia, their form, their possibility of movement.

The first happiness of the child is in becoming acquainted with this autonomous world of obstacles, but he can make this acquaintance only through free activity, through his caprices. The capricious thought is both an initiative and a happenstance, at the same time a negation and an affirmation of the real value of what is encountered. This encounter is transient, because blind expansiveness admits only of itself and excludes the possibility of choice. Through the obstacle offered by things, expansive impulsiveness, as we have said, is reinforced anew, and this time in a highly reactive manner, and it may even become true aggression.

Thus we see that the primary and primitive freedom of caprice is impregnated by another aspect of freedom. It is an acquired freedom, an experienced liberation, and this is identical to the consciousness of victory and the victor. At this moment, the child enters into the *second* phase of his human reality and of his experienced freedom.

He forms his being for himself in his self-consciousness as a being of power. It is the first step on the road where man finds himself as ipseity. But a resolute attitude and an intention of freeing one's self, of conquering, presuppose, to a still higher degree than caprice, that the thing conquered will be encountered as a value – a *value which is worth the effort*. We all still retain the memory of this experience, and can always experience it anew, through the youth which never completely leaves us. Who does not know the joy and the feeling of freedom of the swimmer or the mountain-climber? I refer to such experiences for the precise purpose of disclosing the phenomenal structure of experienced freedom in the consciousness of conquest.

The water through which we swim, or the hill which we climb, which we are going to conquer, are also, it is true, the adversaries of our strong impulsiveness of caprice; however, we also encounter them, through activity, as positive values. In the joy of the act, there is at the same time the joy which the water, the mountain, the space, give us. The valuefulness of that which makes the movement difficult is greater, the more certain the victory. To the poor swimmer, the water is an enemy; to the weary mountain-climber, the mountain is a menace.

But let us turn to the development of the child. The stage which we have described as the second phase of human reality in which the child performs his actions is certainly existence as power, but it is a relative power, rendered *relative* by the situation of our power.

"The idea of situation excludes absolute freedom at the beginning of our activities," says Merleau-Ponty. To this we add: to accept a *relative freedom* is identical with accepting the reality of the world in its own value for us. By this acceptance, the child has the experience of a new characteristic of his freedom. He feels himself free only in his *own* world, the world assumed and projected as his. Things, in this world, are no longer chance obstacles and autonomous objects with respect to a blind expansiveness; they are no longer entirely dependent on the taking of initiative. They appear and are formed in their autonomy, but yet as the property of the subject, as parts of his world.

The child gains, by means of the conqueror's freedom, a relative freedom, and thereby he conquers his own world, in which he confronts a hierarchy of values. Thus the *third* phase of his development marks the appearance of another experience of freedom. To understand this phase correctly, we must understand clearly the modification which relationships undergo. To this end, we must again start from the original dynamism of the child. We have already indicated two basic types of this dynamism, namely, expansive movements and adaptive movements. These latter are endowed with less muscular tension and consequently with less rigidity of articulation, a weaker impulsiveness, and are especially characterized by the enjoyment of tactile sensations. Through his adaptive movements, the child encounters from the very first qualities and forms, that is to say, sensed values.

This valueful world in turn arouses an adaptive motility, modifies the blind expansiveness, and consequently the aggressiveness. Every child is capable of this adaptation to values. But in our traditional culture we consider interest in tactile sensations to be characteristically feminine, and consequently so consider sensitiveness to quality, to form, to color, to sounds, to weight, etc.

Researches in child psychology prove, in fact, that girls are much more capable than boys of executing delicate and subtle movements of the hands, and that they possess a greater sensorial sensitivity.

However, each normal child will acquire, if he receives an appropriate education, the experience of sensed qualities. But he will be able to appreciate them only when he has attained, toward his environment, the attitude which we have indicated as the *third* phase of existence. Certainly this does not mean that the other phases will be liquidated. All the human possibilities are ever present, and it is by virtue of the environment offered to the child by education that one possibility or another will be more or less perfectly realized. Now the existence inaugurated by adaptive dynamism is characterized by the appearance of a double and ambiguous tendency to communicate and to grasp from a distance, tendencies which are equally original.

They are both expressed by an attitude of expectation and attention. This attitude of recoil is the disposition which gives rise to attitudes of surprise, of admiration, and of respect. They are the natural consequences of the adaptive dynamism and they are the condition which must be fulfilled in order to awaken the consciousness of relative freedom.

This latter is identical to *freedom of choice*, which is the prelude to true freedom – moral freedom or freedom of action (G. Marcel). Thus there is manifested in the child, through choice and recognition of values, through surprise and admiration, the true meaning of the "absorbent mind." With G. Madinier,[11] one may call it "a true rational instinct which appears in the first years of life."

11. Gabriel Madinier, *Conscience et amour: essai sur le "nous."* (Paris: Presses Universitaires de France, 1947), p. 75.

How this desire to know and to understand still remains enveloped in random consciousness may be gained from a remark by the same author: "The emotion which corresponds to this 'rational instinct' is surprise, which is nothing other than the principle of sufficient reason or of universal intelligibility, not considered as logical rule but as *vital demand*."

Surprise is in truth the first expression and intimation of the "absorbent mind" which, once formed, will determine the structure of the child's existence during all of his childhood.

It is interesting to note, in our investigation concerning the development of experienced freedom and its relation to moral freedom, that surprise is the source of admiration and the latter the source of respect.

We must distinguish carefully these three intentional attitudes. Surprise signifies the arrest of the consciousness before the opaque and massive factuality of a perceived object. At the same time, this arrest, which we call attention, is impregnated with a presentiment of a possible translucidity. The novelty by which one is surprised must have somewhere an already known trait, should anticipate a certain familiarity.

So far as admiration is concerned, familiarity dominates the situation; it opens the vista to a value which is encountered with a beginning of enjoyment.

This enjoyment is of another character than the joy of the conqueror or the joy accompanying the freedom of caprice. In admiration, the subject rejoices in a disinterested manner; rather, he enjoys aesthetically. He participates in the disclosed value. The objects chosen and admired enlarge his vital space, because the subject assumes the values of his own world, the world in which he is situated.

Through admiration, man enriches himself. For the child, this circumstance is of high importance. Adaptive dynamism can realize only a very few values, but admiration becomes a participation in a universe without limits; it becomes the forgetting of self in the enjoyment of a world of implicit truth and beauty. In this act of fusion, the child exteriorizes himself, and he finds his own value in the admired objects.

He participates in their intelligibility. The consciousness of freedom at the same time undergoes an essential transformation. Experienced freedom becomes the life experience of a freedom, of which "the heart dwells" – according to a profound thought expressed by L. Lavelle,[12] "in an act of acceptance, in a *yes* which we give to being and to life."

Can we call this freedom moral freedom? That depends on the emphasis of the *yes*. If this *yes* means only an affirmation, of the nature of a declaration that a value exists, there is no moral element in admiration. But it is not possible to make precise distinctions in the realm of the relations between the human mind and the world.

Admiration, especially in the child, is often at the same time respect, that is, "the disposition by which the ego recognizes the right of obligation over itself."[13]

12. Louis Lavelle, *Les puissances du moi* (Paris: Flammarion, 1939), p. 51.
13. René Le Senne, "Traité de morale," *loc. cit.*, p. 575.

The child who admires a design, a flower, a bird, is not an aesthete as such, but his intention surpasses pure admiration and is recognized as an obligation to admire and respect beauty, the cleverness of the designer, the mystery of life, the wisdom of the creator.

The unfolding of moral freedom in the child is accomplished in situations which arouse surprise, admiration and respect. This truly human freedom surely draws its vital energy from the experienced freedom of initiative and power. But moral freedom does not reach its full unfolding and its full fecundity until it goes beyond freedom of choice and becomes a creative activity: which implies an act for which the child decides to bear the responsibility. This act is one in which the ego accomplishes and presents itself, in its human dignity, as a personality.

The way in which the three phases of child existence and the three modes of freedom are connected may be illustrated by means of an example.

For this purpose we choose the well known picture of a child who, in a Montessori school, carries a brim-full glass of water to the teacher. Freedom of caprice is suppressed, but the freedom of the conqueror still fills the child's consciousness. At the same time, consciousness is present as the relation of the ego to the person of the teacher, beloved and respected. But it is, in particular, the experience of *objective obligation* which gives to the child experienced evidence of the beginning of his moral freedom.

This objective obligation is identical to the act of comprehending objective demands and assuming them as a responsibility with respect to the situation. It is only through this structure of significance that true moral consciousness is formed, for the obligation proclaims the ego to be double: "*below*, the empirical ego, limited in time and space by the body which incarnates it, a given ego, having a nature resulting from a history; *above*, the ego of value, whose goal is, through obligation, to raise up the first."[14]

The dividing of the ego means only consciousness of a moral conscience and of freedom considered as moral freedom. In our example, the request of the teacher had no other aim or result than that of constructing a situation which, through its objective structure, will be easily assumed by the child as a personal obligation.

In my opinion, one of the most important contributions of Mme. Montessori is the invention of situations of an imperatively obligatory character. Through this educational method, the development of moral freedom no longer depends on the individual orders of the institutor, but is related to a world presented to the child as concrete.

The disclosure of Mme. Montessori rests on an absolute dependence on the value and intelligibility of reality. This reality appeals to the spontaneous dynamism of the child. If this appeal is addressed exclusively to the expansive dynamism, the child will develop only the consciousness of a freedom of initiative and of conquest. If on the other hand the educational situation appeals to the adaptive dynamism – which is done in the Montessori method – freedom is developed in another way. The first two phases, namely freedom of caprice and that of conquest, are thrown

14. *Ibid.*, p. 571.

off at the beginning of their autonomy. They will be integrated into freedom of choice. That happens from the time of the systematic exercising of sensory experiences.

At the beginning the child certainly lets himself be captured by the joy of the sensations. At the same time, consciousness, in the encounter with things, is formed into a consciousness of the valueful existence of the world. At this moment – we may even say at this historic moment – a new chapter begins in the life of the child. It is the *fourth* stage in the relation of experienced freedom to moral freedom.

In this encounter, there is more than an agreeable perception, more than admiration of a value. The "valorized" being becomes a valid state in itself which, in the situation of the encounter, presents itself as a compelling reality.

The child is appealed to, and he must, and he can, respond. By this response, his admiration will be metamorphosized into respect, his aggressiveness transformed into prudence, the things desired transubstantiated into things loved with a disinterested love. This first love of the child for things, plants, animals, comrades, means the introduction into the oral relations of the true unity of society.

"Love appears" – in the child's existence – "as the highest category and the very expression of intelligibility: if we want the world to be intelligible, we must understand it as a work of love."[15]

What the child learns through personal experience in these encounters is that "love is the very end of all things; it depends on itself and justifies itself. It does not consist in wishing well, it is *the Good* itself. It is not good, in fact, because it gives to those which it unites the meaning of exchanging values...." "Values are such because love has produced them in order to fulfill itself."[16]

The profound ideas of G. Madinier resound through Montessori education. This latter gives the child the possibility of transforming, through his own initiative, experienced freedom into moral freedom. This transformation is demanded by the experience of obligation toward all the reality of his environment.

This reality contains, in a form accessible to the child by means of example, all the fundamental categories of the human world: that is, perceptions, thought, nature and culture. While self-consciousness realizes the ego as redoubled by the obligation which is the response given to reality, the child's existence, in its ipseity, is affirmed toward the intelligibility of his world. But there is more: the child in the Montessori school is not isolated. He is in a truly human small society. The ideal reality of this society is union and peace, respect and love. It has been especially impressed upon the teachers to see that the children, unregimented and unconfined, modify their freedom of caprice and of conquest under the effect of the objective power of social relations. Through their more serious cooperative tasks, through mutual aid, and also through the enjoyment of their games, the children build up the "we" as the reality within which each is responsible for the other. Thus is formed, through the experience of social relations, a moral freedom, truly experienced.

15. Madinier, *loc. cit.*, p. 83.
16. *Ibid.*, p. 102.

It seems useful to summarize briefly this ensemble of ideas and to note the result of our phenomenological analysis. We have encountered freedom as the pure climate in which the mind is able to develop to its full perfection. This development receives its principal force from spontaneous life, and herefrom can be manifested, through expansive dynamism, the *first* phase of experienced freedom, as freedom of caprice, with respect to which the world is only an obstacle.

As soon as this freedom takes on the meaning of a subjective resistance, we note a relationship with respect to moral freedom. By the reactive reinforcement of resistance, the primary consciousness of freedom is linked to the secondary freedom of the conqueror. In this *second* stage, the child builds his existence as power and the world as a thing which is *worth the effort*, as a relatively positive value.

Consciousness of activity is relatively free choosing, and signifies the approach of a *third* phase of the development of existence. In this stage, freedom of choice, the prelude of freedom of undertaking, is built on the basis of adaptive dynamism. Little by little the ambiguous attitude of the encounter is generally differentiated into special attitudes of surprise, of admiration, and of respect, through which moral freedom is created in the form of an acceptance of obligation which the concrete world of the child himself demands. The *fourth* stage is attained in the unity of experienced freedom and of moral freedom, in child love, and in the encounter of the "we" as a reality of responsibility.

The development of freedom which I have tried to sketch is not the natural transformation of vital spontaneity into normative conduct. It is only compelling situations which can awaken morality. But a situation can have an obligatory character only for a being already having a normative consciousness to start with.

The most perfect education is that through which integration of nature and mind is best approached and in which, consequently, life is humanized and morality vitalized. The secret of the Montessori method is the perception of the reality of love which is identical with love of reality. The life experience of the unity of this true reality and of this unselfish love is found stated concisely and precisely in the noble thoughts of Romano Guardini,[17] with which I wish to close.

"Love is the attitude which knows that one can possess something as his very own only in the form of a gift.

"He who loves progresses ever farther toward freedom, toward freedom from his own chains, that is, from himself."

27. Romano Guardini, *Welt und Person* (Würzburg: Werkbund-Verlag, 1939), p. 143. Id., *Notizen zu einem Bilde von Dantes Persönlichkeit*, Schildgenossen, 1939, p. 229.

10

The Hotel Room*

D.J. VAN LENNEP

Someone once said: a room is a sacred garment. This expression points to the fact that we dwell in a room not because it is our room, but because it becomes our room when we live in it. What is this mysterious inhabiting through which a room becomes ours? For the most part we believe that it is very important for the inhabiting whether or not, for example, we have placed furniture of our choice in it and that through this choice a room becomes a personal expression of ourself. However the room as an expression of our personality is to a great extent an unfulfillable dream: there are too many factors which do not depend upon our free choice. There even are factors which not at all depend upon our choice, such as size, height, and the distribution of light. As in almost any human situation we must accept what is available. One may say, so far so good; but within the already determined limits there is still room for a choice and for arranging the furniture according to one's own view and taste. But this freedom is limited, also. Bound by limited imagination and by what one received from the parental home and particularly by what is offered by manufacturers and antiquarians as far as new or old furniture is concerned, we set up housekeeping. And even then our freedom remains within a certain range in that without realizing it explicitly we shall make the room correspond to the demands made on our home by a certain rank, class, and culture. For a room is not just merely for ourselves alone, or for our family only, but also for our "social ego," that is to say for society insofar as it is ours: the friends, the acquaintances, whom we receive in our home, co-determine how the home will look in its various rooms. Furthermore, this very class and culture was already expressed in what the shopkeeper had to offer in a certain price-range. Thus it appears that the margin of our personal taste which is freely realizable in the arrangement of our room is so small that if indeed the personality of our room should really depend on this, one would have to ask himself whether he could ever really speak of *our* room. And, indeed, would someone who could afford to build his own house and to have each piece of furniture made according to his own taste be really more at home in his house? We know this is not so and that in such a case one takes the risk of feel-

* "De Hotelkamer" by D.J. van Lennep appeared in *Persoon en wereld: Bijdragen tot de fenomenologische psychologie*. Utrecht: Bijleveld, 1969, pp. 33-40. Reprinted by permission of the Publishers. Translated for this volume by Joseph J. Kockelmans.

ing like a stranger in his own house. However, all of this does not mean that a personal expression in the room-arrangement is completely made impossible for us. Just as is the case with any other human situation, so, too, a room is not personal from the fact that it is a completely original and free act of the will independent of the milieu and the existing possibilities of making or setting up a house; no, the personal element manifests itself in the manner in which one has made and still makes use of it, the manner in which the bought pieces of furniture are used, arranged within the very small margin of possibilities, which a 6 by 9 room for example still leaves us, in the choices of color-scheme for carpeting and pillows, in the manner in which the light is permitted to come in between the parting of the curtains, and particularly in the manner in which all kinds of things which might have a disturbing influence on the dweller have been omitted, and which perhaps without any criticism intended, would have been put down by others. The personal element manifests itself also in the manner in which anything which speaks of tradition, something which has been handed down from a grandmother or a distant cousin and because of this speaks the silent language of familiarity only to the occupant, a language which connects the generations symbolically with one another; thus the manner in which all of this is tolerated or even receives a place of honor in the intimacy of a livingroom or bedroom. The "coziness" of a room is certainly not, or at least not primarily a function of the furniture, but of the manner in which the furniture has been used. The room is very seldom, if ever, impersonal, not even when exclusively impersonal, mass-produced furniture is found in it.

One speaks of the atmosphere of a house, of the "smell" which lingers in a room. It is this smell particularly which has an influence on us, and which can give us the feeling of familiarity or of alienation which means a certain level or class to us. It is difficult to say how a room gets its characteristic smell. It tells us how things and people live in the house, breath, respire and perspire; it emanates from things as well as from people and does not depend on whether or not the house is aired. It is one of the most vital expressions of the life-style of the occupants. But all of this does not yet sufficiently explain the fact that one exists in one's own room as in a garment. We always enter someone's livingroom for the first time with a certain hesitation or embarrassment, that is into the room he "inhabits," not because this room is an expression of himself, but because this dwelling refers to a much more intimate relation than any expression by him could ever be.[1]

What this dwelling precisely is, is difficult to describe when we limit ourselves to considering the dwelling in which we customarily live. We will reach our goal faster if we take our point of departure in particular, and eventually even pathalogical

1. The burglar experiences this too. I am borrowing the following quotation from J.P. Sartre (*Saint Genet* (Paris: Gallimard, 1952), p. 244; English trans. Bernard Frechtman (New York: George Braziller, 1963), p. 260): "If all goes well, one *enters a man*, for the gaping, defenseless apartment, naked and paralyzed, is a man. It reflects a person, his tastes, his ways, his vices: 'I do not think specifically of the proprietor of the place, but all my gestures evoke him... I re-create the absent proprietor. He lives, not facing me, but about me. He is a fluid element which I breathe, which enters me, which inflates my lungs."

forms of dwelling, the kind we experience in the house we dream about, in a hotel room, briefly in the dwelling, with which we are not familiar.

The rooms in our nightly dreams are generally not the dream-rooms about which Bachelard[2] has written, for example the rooms of our youth, the room where mother used to sit.

The rooms in our dreams are different; often they are not familiar. These abstruse rooms which are much too big or much too small and in which we are almost always strangers, rooms with too many doors, rooms from which we run away and which we cannot find again because they are suddenly no longer there, rooms which have half-opened doors and open on halls, rooms we know are empty and which we nonetheless pass full of fear. These are the rooms we often "inhabit" in our sleep, inhabit without being at home in them. Rooms which we enter and leave, in which we lose ourselves and in which we wander as in the labyrinth of our soul. Rooms which we ourselves obviously are, but which nonetheless we are unable to really inhabit. Are these rooms not a "speaking" image of the diffusion, the irrational unspeakableness of our given, co-given, involuntary life? Are they not similar to our body, which we are equally unable to accept completely as our own body and which nevertheless is our own inalienable body?

Sometimes it can happen to us that waking we walk around in our own house with the feeling that the rooms have something ghostly about them, something unreal as in our dream-life. Suddenly we are no longer at home. It seems as if it is the house of a stranger, although there can be no doubt that it is our own house, our own room. This derealization, too, shows that dwelling does not consist exclusively in being in the midst of furniture and carpets which we ourselves have chosen. In such a state the room has lost its familiarity. The pieces of furniture have changed into strangers. They have lost their usual physiognomy. The small old clock from our parental home is just an ordinary timepiece, a cast-off object. Things suddenly stand in meaningless order next to each other. They no longer constitute a real "room." Everything has become useless material, tired and worn-out. And yet we are not present in an anonymous way.

Think how the room in which we feel bored has changed, also. Suddenly we see the fact that all the objects stand in their proper places as no longer real, they are no longer there for us, or to be used by us. But they continue to maintain at least one relation, namely that they affect us as an absence of intimacy. We are certainly not present in an anonymous way in this room, either.

Strictly speaking in our room we are never present in the mode of anonymity, but always in the mode of degrees of or forms of intimacy. This is why when I enter the room of the other, I enter his intimacy.

How little a room is really "expression," and how little expression is related to "dwelling" becomes clear to us the moment we realize that all expressions after a while undergo a process of alienation in regard to ourselves; after some time things are no longer ours. A work of art we made, once it is finished, looks to us as if someone else had made it. Something which at a certain moment expresses our own be-

2. Gaston Bachelard, *La terre et les rêveries du repos* (Paris: Corti, 1948), Chapter 4.

ing shortly afterward no longer does so. However, dwelling transcends our choice as an expression of ourselves; it is the continuous unfolding of ourselves in space because it is our unbroken relation with things surrounding us. It is human existence itself which constitutes space. We simply cannot do otherwise. The things which surround us present themselves in a quality of space which we ourselves are as those who live in this space. The pronoun "my" in the expression "my room" does not express my possession of it, but precisely a relation between me and the room, which means that my spatial existence has come about. We can experience this exceptionally well in the *hotel room*.

My hotel room is my room; for I shall pay for it, perhaps I have even reserved it in advance, but at the same time how little is this room *my* room. The bell-boy leads the way, he takes me to "my" room. I have followed him with some tension and suspicion through certain halls which reminded me of a maze for rats, and suddenly we were standing in front of the door to my room. Now I am standing in this room. My luggage is on the rack destined for that purpose. The bell-boy opens the door of the closet as if he had to convince himself and me that it is empty and checks to see whether the towels are indeed clean. This reminds me from the very first moment that only a few hours before *my* room was someone else's room. Who slept in this bed, washed himself in this sink, used this glass? The room has been made up, the traces carefully removed, the room is fresh and if everything is all right, then there is no "smell." How different is the guestroom of my friends where I am expected, where indeed there is a certain smell, precisely that smell which so vitally represents that house in which I like to be so much.

The hotel room is *for anybody* who can pay for a night's rest, and thus it is for noone. It is a guestroom without a host, unless one wishes to attribute this role to the room clerk, "receptionist" as it is called euphemistically. I am now the *casual* owner of this *accidental* room for a few days and nights. The bed standing here is one of tens or hundreds which have been purchased simultaneously; and the same can be said of almost everything else found in the hotel room. Even the pictures on the walls were bought wholesale. They are pictures which had I owned them myself certainly would have been put in the basement, but in this room I can look at them without indignation. For I do not feel responsible for the fact that they are hanging there. I am here in a completely different mode of existence. I am free from part of my social obligations. I am a stranger, a "number" in a numbered room.

But now, although the idea of the former owner is an always present element when one takes possession of the new hotel room, these feelings disappear quite soon as a function of the process of "inhabiting" the room, a process which takes place almost from the first moment on. Already from the moment I begin to unpack my luggage, I observe that I have started to expand in this room. From being a wanderer in a strange city I have somewhere come to a home of my own. There, that is my closet, here this is my mirror, there that is my window, and that is my bed. I am already "oriented" here. I already inhabit this room. A feeling of security permeates my fatique, I stretch out for a moment on the bed, let my eyes wander over walls and ceiling and quite soon the "expanded" garment has been woven. If in addition I have already slept in this room for one night and the next night return to my hotel tired from my travels, then this room *receives* me, is already *my* room. I come exactly to "number so and so" and am at home in my own home; no other room in this hotel could at this moment be the same thing in the same way. A relation of intimacy has been created between this room and me so that I experience myself as being welcomed while entering it, and this fact alone already proves that I inhabit *this* and not any other room. This feeling already occurs when I insert the key in the key-hole and can indeed turn the lock.

Nowhere can I experience the process of inhabiting as well as precisely in the hotel room because I am here in the midst of wall paper and furniture I did not choose myself, things which in no way are an expression of my personal preference or choice.

But the hotel room gives me not only in a special form the experience of inhabiting, it also teaches me a mode of existing which when I am at home I do not know or only barely know.

In this room for which I do not bear any responsibility, in that it does not indicate my past or my future, in that I merely appear in it as a number in an arbitrary series, I suddenly become freed of my obligations and traditions. I find myself transformed through the anonymity of the hotel room. I experience this for instance when completely or partly undressed I walk through this room, see my image in the full-length mirror of the closet, or distribute my things over the available space in the room. Here I am less directly determined socially in my actions than I am at home where I have to deal with my room as a room which is also for others, a room into which other members of the family have more or less a right to enter and to co-inhabit. Even my study – which is pre-eminently "my" room – is a room belonging to the house, which is essentially "our" house, is a room for which in some respects I am socially responsible. My hotel room on the other hand will be taken care of by an anonymous maid who will "take" everything or at least almost everything and whom perhaps I will never see again. And even though I do not behave differently than I do in my bedroom at home, my actions in my hotel have a freedom, a "looseness," a "being free from everything" which I do not find in this form at home.

If I am alone for one night at home, I cannot treat my bedroom as if it were a hotel without having guilt-feelings. The objects surrounding me speak a language which, if it does not force me, at least asks me to respect the traditional manner of doing things. For this room at home forms a part of my social home even when I am "on my own." The hotel room on the other hand is a non-committal dwelling-place whose use is left completely to me alone. The possession-less possession of the hotel room throws us back upon ourselves in a certain way. Whatever we do we seem to do on the basis of a greater freedom, at least a different type of freedom or a greater voluntariness. This is part of the pleasure we experience in a hotel room when we are on a trip. The hotel room is the room of that which just "happens to us"; it addresses itself to us as an invitation open to all things and multi-valent; free from all historic meaning, from all habit and tradition, duty and obligation it invites us to "what is going to come," to adventure. The hotel room is the room of *adventure*, of whatever is still left of this in our twentieth century. Whoever wishes to learn the deep meaning of adventure must probably go back as far as the 12th century. In the fine commentary which Rato R. Bezzola wrote on *Eric et Enide* by Chrétien de Troyes[3] he describes the 12th-century adventure as the necessary counterpole of the safe home, necessary for the complete development of our human life. It is this adventure which withdraws us from the egocentrism of the family and the egoism-for-two of marital love and gives this love back to the world. This is an adventure in which the woman, too, takes part, albeit in a completely different way from the man. This is the adventure which is simultaneously a conquering of ourselves and of the other, and thus of our love which would fade away between the "walls" of our own life. At the beginning of the adventure we experience an invitation to a new task and however vague this may be, we are plucked from the stable form of life, we are reminded of old and almost faded ideals, we are open to another world, a world with other people, that is to say we are on the threshold of a new social contact; and even the modern hotel room gives us feelings to be experienced which in some sense or other are still connected with this mode of existing. *Each hotel room still indicates something of the adventure*, of adventure in the original sense of *ad-venire*, a meaning which unfortunately to a very great extent has been lost in our time. For as a matter of fact the "adventurer" is the least "romantic."

Who has not experienced this in some way on his honeymoon. The hotel room plays an important role here. Within the framework of the impersonal hotel room one can get to know each other in a way which is impossible for a couple who stays home, and which gives a very special color to this first attempt at living together. The married couple find themselves there in a certain manner in the freedom of mutual discovery. The treasure of the honeymoon experience is later taken along in the married life as lived in the home and deepened; but the discoveries in the hotel room are experienced in a manner which is unique and which offers us a very special framework for the mystery of the first complete encounter. It is a good thing that the first realization of love begins as an adventure, for without the aspect of adventure it is very difficult for love to come to its highest form of development. Marriage itself, as far as its development is concerned, cannot do without this adventure aspect. Many people today know only of the cheap adventure which consists merely in the negativity of domestic order. But marriage itself is the Great Adventure which can exist only if time and again the lovers go beyond the narrow enclosure of dual egoism.

The job of coming home is experienced only by those who have gone through the experiences of an adventure. No adventure without a home, no home without an adventure. The hotel room is not made for those who do not have a home, but precisely for those who possess their own home. And it is particularly the *joy* of the hotel room which, strictly speaking, is only for the married couple or for the genuine love couple for whom the day will bring a new series of discoveries, and not primarily for those who feel that their adventure is already dying at sun-up.

3. R.R. Bezzola, *Le sense de l'aventure et de l'amour* (Paris: Le Jeune Parque, 1947).

Meanwhile, there are many kinds of hotel rooms.

There are those which have lost their habitableness, for instance because of the fact that the furniture is worn-out or neglected. Furniture should refer as little as possible to the use the series of predecessors have made of it. In a "good" hotel room I do not find a soiled spread on my bed, a worn-out carpet full of stains, no wall paper above my pillow filled with grease spots, in other words a room which keeps reminding me of my predecessor. I am one of the so-many in a series of ephemerous visitors who were here before me; a long series of individuals of which I am just one, a tired traveller, a wanderer.

Because of the repugnance which such a room constantly evokes in me I think continuously of my departure. The room is barely or not at all inhabitable, the garment cannot be woven.

The room which is too luxurious, that is the room I cannot really afford appears to be uninhabitable, also. The walls, the closets, the bed, and all the rest refuses to let itself be constituted as my room. I feel like a displaced person in it. The joy of the adventure does not come to full expression. This room reminds me of my financial position and I am aware, also, of the attitude I have chosen to adopt in regard to my own financial problems.

Thus there are many hotel rooms, but all of them have in common the fact that in regard to the temporal occupant they indicate unknown or hidden aspects of his personality, so that we may say it is particularly within this framework of impersonality that the personality comes to an experience of itself.

The good hotel room is impersonal. But for that reason it need not be repelling. It may even be cheerful and cozy. And also completely "inhabitable." This manifests itself when my time to leave has come. When the luggage has been packed and the adventure comes to an end, I have materialized one of my many possibilities of inhabiting a room. It never happens that a man leaves a hotel room which was "inhabited" by him without a certain feeling of sadness. And on the way home I take with me the memories of my many hotel rooms on my journey, memories from which I shall be able to reassume my social position. Enriched and with deeper insight I return from my holiday trip. But this homecoming is as temporal as the trip. When the impressions of the "adventure," of what "has happened" to me on my trip are assimilated, then out of a restless heart new desires for unknown horizons emerge again and new travel plans are made in the intimacy of our own living room. Not without reason has it been said: "... one must preserve in oneself the nostalgia of the country one did not visit; one should never renounce the idea of going on a trip."[4]

4. Julien Green, *Journal*, 4 vols. (Paris: Plon, 1938ff.), vol. III, p. 23.

11

The Psychology of Driving a Car *

D.J. VAN LENNEP

In the pages to follow I shall attempt to consider the activities of the driver from a psychological perspective. In so doing I shall take my point of departure in the ideal driver. Everyone knows that people drive in very different ways and that obviously all of these different ways of driving have their own psychological background. But I shall forego these personal differences and ask what from a psychological point of view can be said about driving a car when the driving is done according to the rules of the game.

First of all we must note that the work-object of the driver is not his *car*, but the *road*. His goal is his destination point, but his field of action is the road. His work is the part of the road to be covered. His instrument is his car which preferably he must know as well as possible and be able to operate. In a modern car the latter has become ever simpler. In America most cars have an automatic transmission. One has even tried to develop a car that can be completely mechanically operated. But however far one proceeds in this direction, that special relationship which exists between the perception of the road and the so-called driver reactions which cannot be made automatic, will never be eliminated. It is precisely in regard to this relationship that important changes in psychological insights have taken place during the past decades. We shall return to this in greater detail shortly. But the question we should be concerned with first is the following: what does a "road" mean, that is to say, how must we define the "road" if one talks of it in a psychological context? If we speak here of "road" we mean not only that long stretch of hardened, brick-paved or cemented earth surface, but the total road-situation with its trees, crossroads, pedestrians, bicycles, drivers, curves, traffic signs, and so on. This total road situation, this ever changing and varying, never completely predictable complex constitutes the special *world* of the driver. The word "world" has a special meaning for the modern psychologist. He understands by the "world" a meaningful whole in which man himself – in this case the driver – is present in a very special way, and with which he is indissolubly connected in a very special manner. In this world the

* "Psychologie van het chaufferen," by D.J. van Lennep appeared in ***Persoon en wereld: Bijdragen tot de fenomenologische psychologie***. Utrecht: Bijleveld, 1969, pp. 155-167. Reprinted by permission of the Publishers. Translated for this volume by Joseph J. Kockelmans.

driver is actively present, not as an observer, not as a spectator, not as a kind of reaction-machine which responds automatically to a field of stimuli, but as an active agent who attributes meaning to the complex of data of which he himself is a part, just as he must accept its objective significations as data.

The perceptual world in which the driver finds himself and which constitutes the object of his activities, is like any other world, a complex of determinate significations. The parts of the whole complex have a completely different meaning for the driver than for the other people who use the road, such as for instance a hiker. For a driver a street-car is a thing which is to be passed on the right, another car a thing to be passed on the left; for the pedestrian on the other hand the street-car is a means of transportation which very seldom happens to cross his path, whereas a car for the pedestrian has the meaning of something which lacks all such steady course and thus remains unpredictable. A car is a vehicle which is to be watched more carefully than the street-car, whereas usually the opposite is true for the driver. A crossing, marked by white lines, is an invitation to the driver to stop and not as it is for the pedestrian, a narrow strip to aid him in reaching the other side of the road in relative safety in the midst of threatening danger. In short, the "what-for" of each thing changes as far as its meaning is concerned depending upon the world to which it belongs. Space and time, too, are unequal for different worlds. The pedestrian lives in another space from the driver: a distance of 500 yards or sometimes even a mile ahead belongs to the space of the driver; the space of the pedestrian is limited to a few yards, or at the most a few tens of yards around him. Often the driver must "count" in fractions of seconds; the time-intervals the pedestrian knows are usually much greater.

All these special data in this special space and special time are ordered in the whole of one single structure.

One often says that the driver must be able to broadly extend his attention; by this one usually means that he must be able to divide his attention among a great number of objects. Regardless of the psychological criticism which can be levelled at these concepts, we feel that even if one exclusively takes into account what is phenomenologically meant here, this representation of things is not in agreement with reality.

What was meant by this so-called broadly extended attention consists in the fact that the natural and objective focal-point of the perceptual field – which in the case of the driver is the road taken in the direction of the longitudinal axis of his car and which is frequently occupied by other road-users – does not coincide with the continuously changing focal-points of his attention: for at one moment he must be attentive on the left, next on the right, then far ahead, now just in front, independent of the question of whether his attention is involuntary, that is, drawn by the objective configuration of his perceptual field, or voluntary, for instance in the case of a crossing where the driver looks for cars which may come from the left or the right.

This representation of things, however, presupposes an antiquated theory of attention and suggests that our attention is a kind of spot-light which we are able to direct toward the things of our surrounding world in order to see them better. This theory separates the function of attention from what, through our attention, is to be seen, as if the things themselves (in harmony with our manner of speaking) would not "command our attention," but would passively and unmoved undergo something like "attention."

In fact, however, attention is the manner in which we relate ourselves to the things on the basis of the meaning they have for us; that is, on the basis of the manner in which they are related to us when we perform a task. Attention is a form of "pregnant" contact. Thus we can never separate attention from our being occupied with the things and from their meaning within our field of action.

And so we must ask the question: what is it that the driver must perceive about the traffic objects in his perceptual field? Well, nothing except their *traffic* meaning in the situation in which he himself finds himself active, that is, their traffic meaning for him.

What do I, for instance, perceive about a car which is parked about 40 yards away? Neither its make, nor its license number, nor whether it is new or old, whether the fenders are dented or not, but only the fact that it is something which I must pass on the left; or even better: that car is for me the signal to observe the road *around it* with a view toward giving way to the left. The car is only marginally conscious in this case, I do not focus on it; its only meaning for me is that it changes that part of the road which is 25 yards in front and in back of it, and 5 yards to the left of it, thus that it changes the context of the road situation. To have perceived this car means to insert myself into the new, changed, traffic situation with the help of certain movements, or as the case may be, with a motor plan anticipating certain movements.

There is no question about my reacting to that car in the sense of a stimulus-response schema. On the contrary I have responded to the entire traffic situation with a completely new insertion of me-in-my-car into the situation.

One can watch everything at the same time. For the driver of a car this is not at all necessary, nor even desirable. However he must have a relatively broad spatial *field* of attention. At a single glance he must survey what is happening on his left and his right and what is happening in front of him farther ahead. But he sees all of this in a determinate structure in which what is important and what is unimportant is divided according to the *meaning* which the things have for him in his situation and which he gives to them on the basis of his own judgment. As is always the case, we have perception only if the data of the perceptual world become structured according to the schema figure and ground. Against the horizon of houses, road-surface, trees, and so on, certain objects force themselves upon him as figures on which his attention becomes focussed; or formulated in a better way with which he keeps up a kind of "interview" which is sometimes short, sometimes long, now dull, now sharp. This holds good for the pedestrian as well as for the driver. But these are other figures than those of the pedestrian on which the driver focusses, and it is another "conversation."

This complex of significations is not static, but changes continuously. What was background a few moments ago, for instance a parked car in the distance, which I did not yet need to watch, becomes figure a few moments later; and conversely, a car which I still have to pass, can change from figure into ground and even before it has been passed belong already completely to the past, as soon as I am certain that "nothing can happen any more." This complex of significations which can stretch out over a large field of attention, does not, however, require a typical "distributive" attention. The complex which is structured according to the schema figure-ground, must be surveyed and judged at a single glance. For the driver there are no loose, isolated elements; all elements belong indissolubly together, they form one structure. A change of one element changes the meaning of another element. But not in the sense that the one, isolated change would have as its consequence another, equally isolated change. Rather the whole field and together with it all of its elements change with the change of one element. It is this field which the driver sees and which invites him to perform activities. I should like to mention here the physiognomy of the road situation, a physiognomy which addresses itself to the driver and to which he responds. But this response is possible only through the traffic rules. The traffic rules structure the field, they make the situation surveyable.

A driver or another road-user who does not obey the rules and uses the road according to his own choosing and freedom, destroys the total structure of the situation, falls outside the norm, his movements become unpredictable for every other road-user. It is precisely because of this, that is, because of the fact that the structure disintegrates, that chances of accidents become great.

Traffic rules have as their function to make possible for us to have a view of the traffic situation in its entirety and this lies precisely in the context of our being involved in the situation. Through the traffic rules all objects which are subject to them, are as it were hidden, woven into a structure which for this reason can be perceived as a whole and to which we can respond accordingly, therefore. Even the driving itself can then become marginal in regard to a kind of spiritual activity, provided the situations are not too complex. However, it can never become unconscious.

Our having just talked about world has still another meaning. A world is always a world with others, a *social* world. There are co-road-users. The other cars which I meet or pass are not simply things, like houses or trees, but things driven by fellow-citizens. The road situation as the object of the driver is pre-eminently a social situation. If I were to find myself merely between "things" on the road, the psychological analysis of driving-a-car would be completely different; and as a matter of fact it does indeed look quite different for a car-racer. The activity of the driver, however solitarily he may sit in the closed case of his car from an outsider's point of view, consists in a piece of social life *par excellence*. For a good driver the other drivers are not traffic obstacles on the way toward his goal which he must try to "take" as skillfully as possible. The normative background of the traffic rules is that the road-users must use the road together and not independently of one another as a group of mad monomaniacs, first of all looking at their speedometers. This also implies that while I am driving I must take completely into account the meaning *my* car has as an object in the traffic field of the *other* drivers.

That this attitude is difficult for the average driver is generally known, but also understandable, for the different drivers indeed all have the same object, "the road," but a different goal which they – precisely because their means toward this goal is a car – try to reach individually. At anytime there is a common goal lacking between men, there is competition, there is a conflict of personal and sometimes even egoistic interests. Thus there is in each driver a continuous battle between two contradictory tendencies. The individual goal must be reached in harmony with others who are striving equally for their individual aims.

Thus the driver must judge the road situation just as one judges the physiognomy of someone with whom one is involved in a conversation. One can read from the intention of the partner this changing physiognomy. A good driver "feels" "where the other wants to go." There is on the road a silent sign-language (of which the turn-signals are the most obvious representatives) which *good* drivers use among themselves. This evaluation takes in the entire road situation, traffic signs included. The interpretation is thus partly in accordance with established arrangements, the traffic rules, partly a matter of immediate intuition. Anyone who is a bad listener is usually a bad driver. One not only has to take into account and thus listen to the very special expressive phenomena of every other driver, but also to the meaning one has oneself for the other drivers. Whoever forgets to turn on the turn-signal forgets that he himself means something in the field of the other.

We have already said previously that perception of the road situation is most closely connected with the operation of the car. We must return to this now here. The driver does not *steer* his car through the traffic situation. This mode of expression would give us the impression that the driver, as it were, stands outside his car and manipulates his car as an object in the midst of other objects. The good driver forms a unity with his car, that is to say his car becomes part of his body. The peculiar aspect of an instrument is that man uses it as an extension of his body. One could say that when we draw a line in the sand with the end of a stick, our touch has been extended to the end of the stick. The driver is as wide as his car. He does not "measure" whether or not he can pass through a space, but "feels" it after a while. But he feels it only insofar and for as long as he sits behind the wheel. Whoever sits beside the driver experiences all the misery of the so-called co-driving. We are frightened of objects which would not have frightened us, if we ourselves had been sitting behind the wheel. The reason for this is that if we know how to drive, we *see* the road situation with *the pedals and the wheel.* Our motor-reactions not only depend (as was thought formerly) on our perception, but our perception, which in our practical life is always a pragmatic perception, equally depends upon our motor involvement. First we see how a situation is for *us*, if we can take positive action. Anyone who after having been a passenger suddenly takes the wheel knows how the aspect of the world changes immediately. The complex of significations is completely different depending upon whether or not we sit beside the wheel or behind it. We experience other things and different aspects of the same things. The figure-ground relation has fundamentally changed. A pretty girl who would have been a "figure" for us if we had been a passenger, is now suddenly background or at least should be so. An open spot in the traffic which while we were sitting beside the driver we were not or at least were only with difficulty able to judge as to the possibility or impossibility of our being able to get through, we can now, that is now that we are sitting behind the wheel with our foot on the pedal, evaluate accurately and without hesitation. Thus the space and time of our world and the meaning of the objects in that world depend upon the possibility we have to behave actively in that world with our entire body, and in our case with a body which has widened itself to include the car. To drive a car is not to be a passive spectator and make decisions accordingly, but to actively participate in a social life which dictates meaning for us and to which we attribute meaning.

But in the second place, in our evaluation of the situation we must derive our judgment from the situations which are still to come and are not yet real. We can express this by saying that we must see ahead in space and time.

In space: for we must not concern ourselves just with what happens in our immediate environment, for in a few moments we will be perhaps hundreds of yards farther ahead, we must look ahead in the literal meaning of the term; but equally in time, for if later we are 500 yeards ahead, this situation will already have changed again, – thus we must evaluate the situation in its *possible* developments, also. Let us not forget here that the word "fore-sight-ly" (*voorzichtig*) is connected with the verb "to see ahead" (*vooruitzien*). A driver who looks ahead is a careful driver. To drive carefully does not necessarily mean to drive slowly, but rather to drive in a way that in evaluating the road situation one takes into account any possible development.

To mention one example, a careful driver will very seldom use his brake over a long distance on an open road because to a very great extent he has already taken into consideration the situation in front of him by the way in which he has regulated his speed. The carefully evaluating-perceiving of the road situation which is possible only when we are actively engaged in the situation with our entire body, is the core of the purely psychological analysis of driving. It is much more important than the so-called quick reaction-time which still lingers as a necessary condition, a remnant of the functional analysis of earlier days.

It is not the quick reaction to an evaluated stimulus which is important in driving, but rather the continuous quick realization of the meaning of the entire road picture. Reaction is a category from an obsolete way of looking at the problem. To realize the meaning of the road situation is to act, it is tantamount to the active participation of the entire man in the traffic situation itself. While I am steering and regulating my speed with my pedal the road situation has for me a meaning which is determinative for my place in the situation.

From this it follows at once that I drive exactly the way I *am*. My style of driving is not determined by my more or less quickly reacting to the stimuli of a so-called outer-world, but by the manner in which I actively take part in the road situation while I observe the established significations of the traffic signs, signals, and so on, and of myself give those meanings to the possibilities which offer themselves to me as an active member of this form of social life; it is *all* of these things which make it possible for me to reach my traffic goal in the correct way. This style of driving which is decisive for my being or not being a good driver, not only involves my "reaction ability" but my entire psychophysical personality.

A third fact – to which we have already pointed, which is of great importance in driving and which because of this is mentioned concretely in almost every traffic accident, is the feeling of responsibility. The careful, evaluating-perceiving of the road situation in the activity of driving must be borne by a never relaxed feeling of responsibility in regard to the other drivers. Traffic is not a game in which one can win or lose depending upon skill and upon having more or less luck. The road situation is part of a social world. The other drivers are fellow-men all of whom have just as much right as I to freedom of action within the limits of the traffic rules. Other cars are not things which move on the road-surface according to strictly determined rules but, as it were, are expanded bodies of fellow-men who have their own freedom of action and therefore, in the last instance, unpredictability. In order to make this degree of unpredictability as minimal as possible, traffic rules have been enacted which are of such a nature that if everyone were to observe them strictly, accidents would be kept to a minimum. Thus one must drive in a communicative connection with the other drivers. This implies a demand for a kind of traffic etiquette. But particularly a demand that I adapt my driving to the degree of my own capability. Not everyone can reach an equally high degree of virtuosity in his careful, evaluating-perceiving. The feeling of responsibility demands that I adapt my speed to the degree of my capability and do not take risks which I cannot handle or which go beyond my driving capacity. A good insight into one's own ability is a strict demand for a driver with a feeling of responsibility.

The practical consequences of the preceding, largely theoretical observations can be summarized as follows:

A careful driver is a man who continually makes decisions on the basis of quick hypotheses about what will happen shortly, in regard to the possible developments of the traffic situation in which he is involved. This taking-into-account of the *total* road-situation has two aspects: the aspect of the traffic situation insofar as it has meaning *for me* (and this we wish to explain in two points, namely 1a) and 1b)) – and the aspect of the traffic situation insofar as it also has meaning for *someone else*.

1a) The most obvious aspect is the fact that one must take into account the traffic situation as this exists for me. I evaluate everything in front and in back of me, the objective state of the road in relation to the properties of my car and my own driving skill, and make that decision which, based on all these data, is the most prudent at the moment; and in so doing I also take into account what presumably will happen a few moments later, but equally with all the other possibilities. That is to say, I must be prepared for the fact that the situation may develop differently than I had anticipated in my quick hypotheses. The other drivers are all human beings each one of whom makes his own decisions, and although I may find it probable that they will behave in harmony with sound reasoning and the traffic rules, I must take into account the imperfection of each driver and the ever present factor of unpredictability in the developing traffic situation.

1b) In so doing I must realize, for instance, that the man in front of me in his turn evaluates the traffic situation from his point of view; I must always more or less try to see the road through his eyes. If he has to pass a traffic obstacle on the left, he will have to move to the left and thus for me this is not the moment to try to pass him.

Just to mention one instructive example from many possible examples where someone did not take into account the *total* traffic situation, may I remind you of the many cases in which an accident has *almost* happened because we neglected to look in the mirror to see how the situation was developing behind us. Everything found in the road situation in front of us was quickly and exactly evaluated and we made our decision in the best of conscience. However we forgot that a good driver must also have "eyes in the back of his head." What is in back of us must be as equally strongly "present" to us as what is in front of us. We must not let ourselves be fascinated with what we directly see, but must relate this continually to what we see indirectly in the mirror. The traffic situation is not something which just files past, but is something in the midst of which we find ourselves at each moment; this means that we are always in the center of a space with a foreground and a background, as well as sides, a space of which we are a part and from which we cannot disengage ourselves. Whoever lets himself be fascinated by what he sees in front of him, can at any moment commit a grave traffic error.

2. In addition to the aspect which we have just discussed, namely that we must take into account what the entire road-situation means for me, there is still another aspect that plays its part, and that is that I must continually take into account what my car means for the other drivers in the traffic situation.

This aspect of the matter is at least as important as the first. I must continually make sure that as a traffic object for the other I am a datum which is as clear and predictable as possible. The traffic situation is always one and indivisible at each moment of the traffic process.

Nothing shows so clearly whether the driver is convinced that driving occurs in a social situation, or whether his attitude, on the other hand, is egocentric, egoistic, or even a-social, as the fact that he does, or does not, take into account what his driving, his car, means for the other. For instance it is possible that as far as oneself is concerned one can have evaluated the situation exactly, but if at the same time one forgets to turn on the turn-signal there is proof that one has neglected a second aspect, or in other words that one does not drive with a clear awareness of what one means for others.

In the analysis of "near-accidents" one can repeatedly conclude to the basic mistake that someone did not take into account the road situation in its totality, obviously something quite different from the unrealizable and incorrect demand that one should pay attention to everything. Driving correctly is tantamount to paying attention to the entire road situation, not only to the road as it has its meaning for me, but also insofar as it means something to the other drivers.

That is why we call good driving a *social matter*, a *social* situation and appeal time and again to the civic spirit of all drivers. We fight egoism, egocentrism and all the other characteristics of people who just let themselves go. From the most recent American investigations into the personality of the accident-prone driver we have received confirmation of the hypothesis that social maturity and thus feeling of responsibility connected with sound reasoning are the best guarantees for safe use of the road. Why then is it that so many people are unable to assume these qualities precisely when they are sitting behind the wheel? Is it possible that in driving itself there is a factor which makes this particularly difficult not just for the dangerous driver but for *all* drivers as well? This question is not entirely new. In his *Psychology of the Driver* Piret[1] states that he believes the joy of driving a car is to be found in an exaltation of the power-instinct. By identifying oneself with the power of the modern motor this power-instinct celebrates dangerous triumphs, particularly among our youth. This is an answer worthy of being studied. But there is no room for the concept of "instinct" in a phenomenological explanation.

Furthermore it is questionable whether one gets much further with an explanation of the power-instinct, although there will certainly be drivers who overcompensate for the lack of personal success in their everyday life by forcing their way in the traffic situation. Taken in itself, however, Piret has correctly seen that the thesis "a man *drives* the way he is" is to be completed by the statement "a man in his car is not just a 'pedestrian' behind the wheel." In other words in many cases a certain *change* does indeed take place in a man who sits behind the wheel. He is *not* exactly the same man he is at home in front of the fireplace.

And in addition to the question of what people are predisposed to causing accidents, one must certainly also ask the question: does the *situation* of the driver, itself, perhaps contain the danger that it changes a man in a certain direction which is disastrous for safety on our highways? It seems indeed that we must answer this question in the affirmative. But this is not, at least not primarily, because driving awakens certain primitive instincts in a man; but rather because the situation of driving itself, and equally so when one takes into account the traffic rules, has more of an *isolationist* influence than a socializing one. And because of this the driver of a car is placed in an extremely ambiguous, psychological situation. For although it may be true that the traffic situation is a "social" one, nonetheless it is of a very special kind. In our descriptions of normal social intercourse we use such expressions as: we come in contact with one another, we approach one another, we walk along with one another, we stay close to one another and are eventually pleasantly surprised by an encounter. But what does one and must one do in high-speed traffic? What terms do we use then? And what is the situation when I sit in my car? Being seated in that cozy small glass room, *separated* from the rest of the world and looking out of the *window*, one expects me to act courteously toward the other cars whose drivers I often cannot see and am even not allowed to see, but whom I must just *imagine*. I must *let these* fellow-citizens *pass* or must *pass* them as politely as possible, or I must remain behind them, or give them the right of way, or even stop. In other words, using the road together always implies something *negative* in regard to contact; gestures people make on the road do not easily remove the *isolation* for each driver because they accentuate it at the same time. We are seldom *pleasantly surprised* by another driver, usually *hindered*. On the road I come in contact with a fellow-driver in order to *get rid of* him again as quickly as possible,

1. Roger Piret, *Psychologie de l'automobiliste et sécurité routière* (Liège: Desoer, 1952). This is an excellent monograph with an extensive bibliography.

albeit in the most correct and polite manner, but at any rate in order *not* to come in *closer* "contact" with him. In normal social intercourse every new contact contains the promise of a possible *closer* acquaintance and contact, but this is precisely what we wish to avoid on the road. It is even a demand of traffic safety that I continuously keep *distance*. The "conversation" which I undertake with my fellow-driver is certainly the slightest amount of "engaging."

The polite traffic situation, which we usually properly call a function of a developed social sense, simultaneously accentuates the isolationist aspect of driving as such which gives something contradictory, ambivalent, to the entire traffic situation, and thus something which I wish to call *irrealizing*, that is to say something because of which driving as a situation can never completely reach the reality of our social life, at least not automatically. To use a paradoxical expression: it is keeping contact by avoiding contact.

That is perhaps the reason, also, that even as adult men we can amuse ourselves so immensely in the minature cars at a fair. There we try to come in contact with one another precisely in order to materialize it, and in these small cars one sees that there is something like a "dream" of the collision which is experienced with genuine exaltation. And one should particularly not confuse this with a form of aggression.

Furthermore, for many people who are not yet adapted to modern technology, the great speed which is reached by an activity involving such little effort as a push on the gas pedal, implies in itself the danger of a certain kind of irrealization. Piret correctly said that "The great danger of the actual car is to be found in its easiness."[2]

For, indeed, a walker who wishes to walk faster must exert his *whole* body; he is completely active. But in our car we are active and passive at the same time. Particularly passive. We say we are "driving," but the truth is that we are equally driven, although it is our duty to be actively involved in the traffic situation. But proportionately it makes no difference to my bodily effort (namely the push on the pedal) whether I drive 15 miles per hour or 65 miles per hour. There is at most a mental difference: anything too easy derealizes a man, makes him, as it were, experience his world as a dream-world. The landscape I cross at a speed of 60 miles per hour is changed, is less real, or at least of an unusual "reality." On the other hand, driving easily heightens our feeling of *freedom*, which – in that it is connected with separateness and effortless arbitrariness – has an irreal aspect, also.

The car is the realization of one of man's dreams of greater freedom; but whoever lets himself be carried away in this dream is prone to permit himself liberties which lead to catastrophes in reality. J.-P. Sartre once said sarcastically: "The Americans believe they are free in their cars."

But this is true for *every* driver to a greater or lesser extent. And it is precisely here that all the dangers of a-social behavior are to be found. For safe driving demands our being engaged in a *difficult* reality: the difficult and ambiguous reality of our driving with others and yet our being isolated in the situation of one road.

2. *Ibid.*, p. *iii.*

Thus it is contained in the driving situation itself, that the driver becomes isolated from the social sphere in which he normally lives with his fellow-citizens. And yet he is expected here to behave socially, but in a manner which precisely accentuates his isolation. It is obvious that egoistic, egocentric characters are particularly apt to fall into the trap of this isolating process. Perhaps it is precisely this irrealizing process which unconsciously attracts a great number of drivers. And it is very likely that high-speed traffic without accidents will never be possible. *In his car a man a-socializes himself.*

However, it remains our duty as drivers that we – fully aware of our situation – observe the game-rules of this wonderful living-together on the road.

12

The Meaning of Being-Ill*

J.H. VAN DEN BERG

If we wish to realize what it means to be ill, we should begin by trying to remember what happened when a short, passing, but positively uncomfortable, illness made us stay in bed for a few days. What will come back to us might be something like the following report.

REPORT BY THE FATHER OF A FAMILY

After a restless and disturbed sleep, I wake up in the morning, not feeling too well. I get out of bed, however, intending to start the day in the usual manner. But soon I notice that I cannot. I have a headache; I feel sick. I notice an uncontrollable urge to vomit and I deem myself so incapable of facing the day that I convince myself that I am ill. I return to the bed I just left with every intention of staying there for a while. The thermometer shows that my decision was not unreasonable. My wife's cautious inquiry whether I would like something for breakfast makes the reason much clearer. I am *really* ill. I give up my coffee and toast, as I give up everything the day was to bring, all the plans and the duties. And to prove that I am abandoning these completely I turn to the wall, nestle myself in my bed, which guarantees a comparative well-being by its warm invitation to passivity, and close my eyes. But I find that I cannot sleep.

Then, slowly, but irrevocably, a change, characteristic of the sickbed, establishes itself. I hear the day begin. From downstairs the sounds of household activities penetrate into the bedroom. The children are called for breakfast. Loud hasty voices are evidence of the fact that their owners have to go to school in a few minutes. A handkerchief has to be found, and a bookbag. Quick young legs run up and down the stairs. How familiar, and at the same time how utterly strange things are; how near and yet how far away they are. What I am hearing is the beginning of my daily existence, with this difference, though, that now I have no function in

* This essay appeared in Jan H. van den Berg, *The Psychology of the Sickbed*. Duquesne University Press, 1966, pp. 23-50 and 61-74. *The Psychology of the Sickbed* is an English translation of *Psychologie van het ziekbed*. Nijkerk: G.F. Callenbach BV. Reprinted by permission of the Publisher of Callenbach and Duquesne University Press.

it. In a way I still belong completely to what happens downstairs; I take a share in the noises I hear, but at the same time everything passes me by, everything happens at a great distance. "Is Daddy ill?" a voice calls out; even at this early moment, it has ceased to consider that I can hear it. "Yes, Daddy is ill." A moment later the door opens and they come to say goodbye. They remain just as remote. The distance I measured in the sounds from downstairs appears even greater, if possible now that they are at my bedside, with their fresh clean faces and lively gestures. Everything about them indicates the normal healthy day, the day of work and play, of street and school. The day outside the house, in which "outside" has acquired a new special meaning for me, a meaning emphasizing my exclusion.

I hear that the day has begun out in the street. It makes itself heard; cars pull away and blow their horns, and boys shout to one another. I have not heard the sounds of the street like this for years, from such an enormous distance. The doorbell rings; it is the milkman, the postman, or an acquaintance; whoever it is I have nothing to do with him. The telephone rings; for a moment I try to be interested enough to listen, but again I soon submit to the inevitable, reassuring, but at the same time slightly discouraging, knowledge that I have to relinquish everything. I have ceased to belong; I have no part in it.

The world has shrunk to the size of my bedroom, or rather my bed. For even if I set foot on the floor it seems as if I am entering a *terra incognita*. Going to the bathroom is an unfriendly, slightly unreal, excursion. With the feeling of coming home I pull the blankets over me. The horizon is narrowed to the edge of my bed and even this bed is not completely my domain. Apart from where I am lying it is cold and uncomfortable; the pillow only welcomes me where my head touches it. Every move is a small conquest.

CHANGE OF THE FUTURE AND THE PAST

The horizon in time, too, is narrowed. The plans of yesterday lose their meaning and their importance; they have hardly any real value. They seem more complicated, more exhausting, more foolish and ambitious than I saw them the day before. All that awaits me becomes tasteless, or even distasteful. The past seems saturated with trivialities. It appears to me that I hardly ever tackled my real tasks. Future and past lose their outlines; I withdraw from both and I live in the confined present of this bed which guards me against the things that were and those that will be. Under normal circumstances I live in the future, and in the past as far as the future draws upon it to prescribe my duties. Apart from a few special moments I never really live in the present, I never think of it. But the sickbed does not allow me to escape from the present.

Normally I am not aware of my body; it performs its tasks like an instrument. Now that I am ill, I become acutely aware of a bodily existence, which makes itself felt in a general malaise, in a dull headache and in a vague nausea. The body which used to be a condition becomes the sole content of the moment. The present, while always serving the future, and therefore often being an effect of the past, becomes saturated with itself. As a patient I live with a useless body in a disconnected present.

Everything gets an "actual" meaning, and this is quite a discovery for us who are pledged to the future. The telephone, rather than conveying the message from the person at the other end of the line, makes me aware of the fact that, as a frozen appeal, it rings with a new sound through a house which has become remarkably remote and strange. The blankets of my bed, articles so much devoted to utility that they used to disappear behind the goal they served, so that in my normal condition I could not possibly have said what color they are, become jungles of colored threads in which my eye laboriously finds its way. The sheets are immeasurable white plains with deep crevasses, steep slopes and insurmountable summits; a polar landscape to the paralyzed traveller that I am.

The wallpaper which I only noticed vaguely, if I ever saw it at all, has to be painfully analyzed in lines, dots, smaller and larger figures. I feel an urge to examine the symmetrical pattern, and to see in it caricatures of people, animals and things. It is as if I am taking a Rorschach-test, immensely enlarged. Hopeless and nightmarish interpretations urge themselves upon me, particularly when I am running a fever. And I feel I am going mad when I find a spot that cannot be made to fit into the structure which took me such pains to evolve.

After a few days I begin to hate the oil painting on the wall. For by this time I have acquired a certain freedom to change the caricatures of the wallpaper; I can replace the configuration I created by another one when I am bored with it. But the figures in the painting, the people, the animals, the houses and the trees, resist every attempt in this direction. The hunter, about to shoot the flying duck, remains aiming motionlessly, while I have judged his chances a hundred times. And the duck, which would probably manage to reach a hiding place if it is quick enough, defies all dangers as it comfortably floats over the landscape where the sun forgets the laws of cosmography in an eternal sunset. "Oh! please, hurry up" I say, exasperated, and even if I am amused at my own words, I do ask the next visitor to please be kind enough either to turn the picture to the wall or to remove it altogether.

THE CALL OF THINGS

As I notice my clothes, hanging over the chair at the foot of my bed, I realize with a new clarity that the horizon of my existence is narrowed. For the jacket there, the shirt and tie, belong to the outside world. I see myself descending the stairs, going to work, and receiving guests. Cèrtainly I am that man, but at the same time I have ceased to be he. The clothes are completely familiar and very near, and yet they belong just as truly to a world which is no longer mine. I feel a vague sympathy for these clothes, which remind me, tactfully, of my healthy existence, which must have had its value. Nevertheless, I am pleased when caring hands change my bedroom into a proper sickroom and my clothes are put away in the wardrobe. For however tactful the reminder is, I do not like to be reminded at all. After all, I cannot and will not put it into effect anyway.

If I am sensitive this way, if I possess the remarkable sense which enables people to understand the language of the lifeless objects, the discovery of my shoes is particularly revealing, even if I find it hard to put into words what these shoes, with their silent and yet expressive faces, have to say. In his famous journal,[1] Julian Green drew our attention to the fact that it is the hat and shoes that are the most personal of our clothes. None of our clothes is entirely anonymous; they are all part of ourselves in a way, an extra skin, the skin that we choose to show others and which we want to see ourselves. We choose our various articles of clothing with this showing and seeing in mind.

A man has not very much choice in this respect. A suit is a suit; the colors may vary a bit, the material and the cut may depend on the amount of money he can afford or is willing to spend. But that is all the variety at his disposal. A man who respects himself buys a shirt that hardly differs from the one his neighbor or colleague wears. In the matter of ties we are less restricted. The salesman shows us a rainbow of colors and an array of designs. A tie can be a very personal thing. That is why we are not really pleased to find another man wearing the same tie; it seems as if we meet an attribute of ourselves which he has unlawfully appropriated.

And then the hat. Even for men the varieties in color, shape, consistency, hairiness and handiness are almost inexhaustible. It becomes even more personal when the first newness has worn off. The hat acquires dents and creases; the brim gets a twist and a wave. These things are all signatures of the wearer and show his hold on things, his way of life. There are crying hats, proud hats, provocative hats, gloomy hats, tortured hats. And just as they tell us something about their respective owners, they certainly have something to say to their wearers themselves. Will the owner of the gloomy hat not be touched with a certain pity when he sees his hat hanging among happier members of its kind.

Shoes, too, form a very personal part of our clothing. Besides that, they enjoy the extraordinary privilege of having faces. Some shoes shake with laughter; others stare silently upon a vague distance; others again look at us full of reproach. In a store we cannot see these things yet; in their distinctive neutrality they make our choice difficult. But we have only to wear our new acquisitions a few weeks and the personality is there. As a rule their faces are not unlike those of their wardrobe-mates. After all, they are of one family. Our shoes constitute our contact with the earth; they tread on country lanes and city streets. Their route is our life's course. Now they are waiting for us, there, by the bed, a silent but futile invitation. The faces with which they look at me completely explain my condition: I no longer belong to the life which none the less is still mine; my street, my road, lies outside the horizon of my existence.

These are a few experiences of a short and harmless illness. Let us see now how a person is affected whose illness is not an incidental intermission, but a long, or even lasting, condition.

1. Julian Green, *Journal II* (1935-1939) (Paris: Plon, 1939), p. 232.

STEVENSON'S ILL MAN

In his captivating novel *The Bottle Imp* Robert Louis Stevenson tells the story of a man with whom all went well in his life. With the help of a magical power, which lives in a bottle, he has become rich. He buys himself a wonderful house on one of the sunny islands of the Pacific. He has it furnished to his taste, sparing neither money nor trouble. And he marries a beautiful and charming girl who fits exactly into these surroundings. When he wakes up in the morning he sings as he gets out of bed, and singing, he washes his healthy body. On a certain morning his wife hears the singing suddenly stop. Surprised by the silence she goes to investigate. She discovers her husband in a state of silent consternation. As an explanation he points at a small insignificant pale spot on his body. He has leprosy. At the discovery of this seemingly insignificant change, his whole existence is ruined. It is no longer of any interest to him that he is a rich man, the owner of one of the most wonderful houses in the world. No longer has he an eye for the beauty of his island; this beauty has disappeared; at the most it is an accentuation of his despair. If he thought of the happiness of his marriage just a moment ago, now his wife belongs to the caste of the healthy, inaccessible to him from now on.

DISCOVERY OF A SERIOUS ILLNESS

Every year thousands of people make a similar discovery. The woman who, while taking a bath, feels "the lump in her breast" puts away the soap and tells herself the terrifying message that death has entered her existence. The man who suffered from constipation for a few months and who hears the doctor say that a very serious operation is necessary sees the scenes of his life change into other ones – and what a difference they make. Less disastrous but no less extensive are the considerations of the student who is told that he has to spend "a few months" in a sanatorium. And those of a housewife who hears that the condition of her heart makes it necessary for her to employ permanent help and that if she does not leave her bed and her easy chair, she may "live for years."

The beginning of every serious illness is a halt. Normal life is at an end. Another life takes its place, a life of a completely unknown nature. Although the patient may anxiously anticipate its sorrows and its sufferings, he never completely realizes what they will be. He considers this life of such low value that he cannot give it a form of its own. It seems to him that it cannot be really lived; it has to be passively endured. The patient feels beaten, bewildered, rebellious or – which is just as unfruitful – resigned. It is an experience of complete surprise, hardly imaginable to a healthy person. One suddenly becomes uncertain about things taken most for granted: the personal function in the existence of others, the necessity or even the indispensability of this function, the faith in a still unpredictable future and the faith in the integrity of the body – even if he theoretically accepts these and similar certainties, they are never *really* certain for him.

And just because access to normal healthy life is barred, this life urges itself upon him with a new promise. The world grows dearer to him than ever. The little daily matters seem more desirable to him than ever before. When Bernanos' village priest[2] knows that he is suffering from an inoperable carcinoma and that his days are numbered, the world flows away from him very rapidly – the experience of lonesome isolation within a narrow horizon – but, at the same time, the world acquires a colorfulness and intimacy that amazes him. He had always believed that the attractions of earthly life had long ceased to affect him. With an incredible clarity, he remembers his healthy existence. And it is certainly not just a coincidence that the first thing that comes to his mind are the roads on which his feet trod; the road that led home from school when he was a child; the hot *route nationale* in summer and the muddy endless country lanes of his parish and finally the road he sees before him. These are suddenly an urgent incentive to live and work. The certainty of death makes life very much alive. The certainty of a chronic illness stresses the blessings of health. The astonishment felt by the patient at this discovery never really disappears as long as his illness lasts.

CONFLICT WITH SURROUNDINGS

Apart from this conflict with himself there is another, possibly even more distressing, "conflict with his surroundings": *Le drame avec l'entourage*, says France Pastorelli. In her autobiographical notes on the sickbed of a heart patient she gave us an excellent description of the state of being ill, from which we have borrowed freely.[3] The conflict with the environment is largely caused by the behavior of the healthy visitors. Whereas the patient never really gets over the astonishment of the beginning of his illness, and the sickbed is never an accepted fact but always an emergency, which, even if it lasts for years and never really holds a promise for recovery, keeps its temporary and incidental character, the healthy visitor quickly removes him from the world of his daily existence and accepts him as a *fait accompli*. Even after the first week the visitors have ceased to look surprised as they enter the sickroom, while it is just this surprise that demonstrates the contents of the patient's life. Even after a month their inquiry "How is everything?" is just as conventional as the gestures by which they put their hats on the bed and their coats over a chair. They barely expect a reply and the patient cannot fail to notice it. If he tries to explain how things are, how those things are that keep him anxiously alert and which hurt and torture him, he cannot fail to see how his words hardly reach his visitors; he can see them think of other things while trying to be interested. He can see them think: "We know, we have known it for a quite a while, and we know it for the future; our knowledge even includes the slight chances of recovery or the certainty of a fatal end."

The sickbed soon becomes a fact for the visitor whereas it never really becomes a fact for the patient. The way in which the visitor talks about the life which only

2. Georges Bernanos, *Journal d'un curé de campagne* (Paris: Plon, 1936), p. 336.
3. France Pastorelli, *Service et grandeur de la maladie* (Paris: Plon, 1933).

just a short time ago used to be his own reality, proves to the patient that he has simply ceased to be part of that life. He has become a complete stranger to the visitor, a non-participant, "just a patient." Everywhere his place is taken by others. While during the first few days or weeks it seemed that his absence would create problems, now these problems are no longer apparent. No one seems to worry about them. Things are going pretty well without him, he is not needed anymore. So little do people miss him that his absence is not even noticed. They only remember him when the calendar says that it is time to pay the patient another visit.

Of course, all this is never expressed openly; it would even be doing the visitor an injustice to assume that these things occur to him at all. He does not think and speak without sympathy; he has an honestly sympathetic interest. Yet he cannot possibly stop the patient, who is extraordinarily sensitive in this respect, from looking through his words at the harsh fact which brought the visitor to his sickroom in the first place: the fact that he is a "patient," a patient in his sickroom, outside of which life – with an "I have ceased to know you" – goes on regularly and sensibly. Whereas the noises from the street, the noises in the house, the new way in which the light filters through the windows in the morning and in which it is superseded by artificial light in the evening, the new appearance of his room with the slightly too large bouquet of flowers and the slightly too expensive basket of fruit, whereas all these things teach him that life outside his small existence has become strangely foreign, his visitors transform this strangeness into a hostile distance, if only by their coming and by their way of speaking.

This is particularly true when they are being tactful. For what else does this mean to the patient than that he has apparently become a person who has to be approached with tact, cautiously, which means, with a certain distance. He belongs to an existence having new norms, where the things that are said and the things that are not mentioned follow other rules than in the world outside. It is largely the visitor who makes this clear to him – he even urges these new norms upon him. The patient finds this particularly true when he wants to speak about his illness.

JUSTIFIED AND UNJUSTIFIED RESERVE

The visitor cannot be blamed for taking a reserved attitude towards the patient's accounts of his illness. He has a right to refuse to be involved in the particulars of this specific disease, or in the experience of the medical examination and of the treatment. It is a pity that this right is so seldom utilized. For this kind of reserve hardly hurts the patient. But he is certainly hurt when his wish to discuss the state of being ill as a form of life which is hard or impossible to bear, is declined with a trivial remark, or when his urgent questions about life and death are treated with a false optimism, as meaningless to the visitor as it is to him. Many patients suffer because they cannot discuss the problems that really matter to them. It is the healthy person who is to blame for this. It is he who goes through life with a completely unjustifiable levity in these matters.

THE CONTEMPORARY REJECTION OF ILLNESS AND DEATH

We all fear death and illness. But we do not talk about it, neither to others nor to ourselves. Instead of overcoming this fear by meeting sickness and death openly as the most real possibility of our existence, instead of making contact with it in a serious discussion, we escape from this discussion by acting as if illness and death did not exist.

The present social customs make this escape very easy. Reading Huizinga's *The Waning of the Middle Ages*, we learn that in medieval times sickness and death were visible for everyone.[4] The sick walked in the streets; they sat at the side of the road. The lepers made their presence known by sounding their rattle; their processions were a visible warning of the silent processions of plague and cholera, which harassed Europe with a certain regularity. Death had an appearance which no one could ignore. If a person died, the tolling of the bells made the whole town join in his burial. He was carried to his grave, which was dug in the heart of the town, at the graveyard situated around the church. Death was present all the time and so was sickness.

This cannot possibly be said of our time. The sick are removed from our everyday life. They live in hospitals or institutions. A visitor to these centers of sickness usually notices very little of disease and death. There is singing and laughter; there is hardly any suffering to be seen. For serious patients are moved to private rooms; they are more or less sick in secret. When death arrives the other patients do not often notice anything immediately. But after a while, there are a few whispers. Death steals over the ward like a secret; it should never be seen. Occasionally a dying patient is allowed to go home shortly before the end. While the idea of letting a patient return to his family and his trusted environment for a last farewell is certainly laudable, it also means that death is banned from the hospital. Death is not allowed at this place of healing.

And even for the patient himself, death has to be camouflaged. Morphine is administered to him – of course, only to remove the pain. But the universal secret wish that death should come to a patient while he is unconscious is gratified all the same. The graveyard is no longer situated in the center of the town in most countries. Particularly in the larger cities one has to make a search to find the dead. Their last resting place is outside the city, covered by a dome of green, hidden, and out of the way. Like the camouflaged municipal garbage dump – and sometimes near to it – the graveyard gives the impression of a *pleasant garden*. The passer-by suspects a magnificent country *mansion* rather than the realm of death.

4. Johan Huizinga, *The Waning of the Middles Ages*, trans. F. Hopman (London: E Arnold, 1924), passim.

THE RIGHT AND THE WRONG ASPECTS OF THE REJECTION OF ILLNESS AND DEATH

Of course, much is gained with all this. It is hygienically correct that the sick should no longer roam through town and country. It is a good thing in many respects that the village no longer has its idiot. It is only humane to give the sick the care they need: it would be heartless to keep narcotic drugs from them if the end is painful, too painful.

But much is threatened to be lost. Psychologically, it is extremely dangerous to abolish sickness and death from our daily life. Psychologically, it is far from hygienic to remove the insane from our everyday existence. Again psychologically, it cannot be considered altogether a good thing that modern civilized man lacks every contact with decay and refuse, thanks to ever more perfect techniques of waste-removal. Wherever these things happen, the perishableness of our existence – actually the first human reality – becomes a hidden and therefore a much more dangerous menace. Sickness and death become catastrophes. To a person who is completely unprepared, they always come as a surprise.

Modern man can be compared to the young Buddha who, being kept away from every human sorrow by his educators, became exceptionally sensitive to everything that did not fit in his artificial paradise. It is certainly not just a coincidence that in our time so much is being thought and written about anxiety and dread. We have ceased to live with the realities of our existence which, just because of this negation, urge themselves upon us in the form of a vague dread. Dread is at the bottom of our seemingly happy and healthy life.

Bibliography

I. HISTORICAL AND THEORETICAL WORKS

Berg, J.H. van den, *The Phenomenological Approach to Psychiatry. An Introduction to Recent Phenomenological Psychopathology* (Springfield, Illinois: Charles C. Thomas, 1955)

Binswanger, Ludwig, "Existential Analysis, Psychiatry, Schizophrenia," in *Journal of Existential Psychiatry*, 1(1960), pp. 157-175

——, "On the Relationship between Husserl's Phenomenology and Psychological Insight," in *Philosophy and Phenomenological Research*, 2(1941-2), pp. 199-210

Buytendijk, F.J.J., "Die Bedeutung der Phänomenologie Husserls für die Psychologie der Gegenwart," in H.L. Van Breda and J. Taminiaux, eds., *Husserl et la pensée moderne.* Phaenomenologica, vol. 2 (The Hague: Martinus Nijhoff, 1959), pp. 78-114

Drüe, Hermann, *Edmund Husserls System der phänomenologischen Psychologie* (Berlin: Walter de Gruyter & Co., 1963)

Giorgi, Amedeo, *Psychology as a Human Science: A Phenomenologically Based Approach* (New York: Harper and Row, 1970)

Gurwitsch, Aron, *Studies in Phenomenology and Psychology* (Evanston: Northwestern University Press, 1966)

Heidegger, Martin, *Being and Time*, trans. John Macquarrie and Edward Robinson (London: SCM Press, 1962)

Husserl, Edmund, *Cartesian Meditations. An Introduction to Phenomenology*, trans. Dorian Cairns (The Hague: Martinus Nijhoff, 1960)

——, *Phänomenologische Psychologie* (1925). Herausgegeben von Walter Biemel. Husserliana, Band IX (The Hague: Martinus Nijhoff, 1962)

——, *The Crisis of European Sciences and Transcendental Phenomenology. An Introduction to Phenomenological Philosophy*, trans. David Carr (Evanston: Northwestern University Press, 1970)

Kockelmans, Joseph J., *Edmund Husserl's Phenomenological Psychology. An Historico-Critical Study* (Pittsburgh: Duquesne University Press, 1967)

Lefebre, Ludwig, "Die Psychologie von Karl Japsers," in Schilpp, Paul A., *Karl Jaspers* (Stuttgart: Kohlhammer, 1957)

Linschoten, Jan, "Die Unumgänglichkeit der Phänomenologie," in *Jahrbuch für Psychologie und Psychotherapie*, 10(1963), pp. 177-185

——, *Idolen van de psycholoog* (Utrecht: Erven J. Bijleveld, 1964)

——, *Op weg naar een fenomenologische psychologie* (Utrecht: Erven J. Bijleveld, 1959)

Scheler, Max, *Die transzendentale und die psychologische Methode* (Leipzig: G. Thieme, 1920)

Stein, Edith, "Beiträge zur philosophischen Begründung der Psychologie und der Geisteswissenschaften," in *Jahrbuch für Philosophie und phänomenologische Forschung*, 5(1922), pp. 1-116

Strasser, Stephan, *Phenomenology and the Human Sciences* (Pittsburgh: Duquesne University Press, 1963)

II. PUBLICATIONS IN WHICH THE PHENOMENOLOGICAL METHOD IS APPLIED

Binswanger, Ludwig, *Ausgewählte Vorträge und Aufsätze* (Bern: Francke Verlag, 1947)

——, *Grundformen und Erkenntnis menschlichen Daseins* (Zurich: Max Niehans, 1942)

Bachelard, Gaston, *L'air et les songes* (Paris: Corti, 1943)

——, *La psychanalyse du feu* (Paris: Gallimard, 1949)

——, *La terre et les rêveries du repos* (Paris: Corti, 1949)

——, *La terre et les rêveries de la volontê* (Paris: Corti, 1948)

——, *L'eau et les rêves* (Paris: Corti, 1947)

——, *The Poetics of Space*, trans. Maria Jolas (New York: The Orion Press, 1964)

Beets, Nic., *Over lichaamsbeweging en sexualiteit in de puberteit* (Utrecht: Erven J. Bijleveld, 1959)

Berg, J.H. van den, and Linschoten, Jan. eds., *Persoon en wereld: Bijdragen tot de phaenomenologische psychologie* (Utrecht: Erven J. Bijleveld, 1963)

Berg, J.H. van den, *The Psychology of the Sickbed* (Pittsburgh: Duquesne University Press, 1966)

Boss, Medard, *Psychoanalysis and Daseinsanalysis* (New York: Basic Books, 1963)

Buytendijk, F.J.J., *Algemene theorie der menselijke houding en beweging* (Antwerpen: Standaard Boekhandel, 1948)

——, *Attitudes et mouvements: Étude fonctionelle du mouvement humain* (Paris: Desclée de Brouwer, 1957)

——, *De psychologie van de roman* (Utrecht: Het Spectrum, 1950)

——, *De vrouw: Haar natuur, verschijning en bestaan* (Utrecht: Het Spectrum, 1967)

Buytendijk, F.J.J. and Plessner, H., "Die Bedeutung des mimischen Ausdrucks," in *Philosophischer Anzeiger*, 1(1925-6), pp. 72-126

Buytendijk, F.J.J., "Het voetballen," in *Tijdschrift voor Philosophie*, 13(1951), pp. 391-419

——, *Pain: Its Modes and Functions*, trans. Eda O'Shiel (Chicago: The University of Chicago Press, 1962)

——, "The Phenomenological Approach to the Problem of Feelings and Emotion," in M.L. Reymert, ed., *Feelings and Emotions* (New York: McGraw-Hill, 1950)

——, *Traité de psychologie animale* (Paris: Presses Universitaires de France, 1952)

——, *Wesen und Sinn des Spiels* (Berlin: Wolff, 1933)

Graumann, Carl, *Grundlagen einer Phänomenologie und Psychologie der Perspektivität* (Berlin: Walter de Gruyter & Co., 1960)

Gurwitsch, Aron, *The Field of Consciousness* (Pittsburgh: Duquesne University Press, 1964)

Jaspers, Karl, *General Psychopathology*, trans. J. Hoenig and Marian W. Hamilton (Chicago: The University of Chicago Press, 1964)

Linschoten, Jan, "Algemene functieleer," in Ph. Kohnstamm, ed., *Inleiding in de psychologie* (Groningen: Wolters, 1955), pp. 50-203

——, "Aspecten der sexuele incarnatie," in J.H. van den Berg and J. Linschoten, eds., *Persoon en wereld* (Utrecht: Erven J. Bijleveld, 1956), pp. 74-126

——, "De psychologie van de sport" (with F.J.J. Buytendijk), in J. Jongbloed and J. Jongh, eds., *Sportgeneeskunde* (Utrecht: Oosthoek's Uitgeversmaatschappij, 1955), pp. 79-88

——, "Dieses Schwebende, das seltsam uns angeht": Fragmente zur Phänomenologie der Inkarnation," in *Rencontre / Encounter / Begegnung.* Contributions à une psychologie humaine dédiées au Professeur F.J.J. Buytendijk (Utrecht: Het Spectrum, 1957), pp. 260-283.

——, "Die Strasse und die unendliche Ferne," in *Situation*, 1(1954), pp. 236-260

——, "Logische en phaenomenologische analyse der bewegingsverschijnselen," in *Tijdschrift voor Philosophie*, 12(1950), pp. 668-728

——, "Over de humor," in *Tijdschrift voor Philosophie*, 13(1951), pp. 603-666

——, "Over het Inslapen," in *Tijdschrift voor Philosophie*, 14(1952), pp. 207-264

May, Rollo, Ernest Angel, and Ellenberger, Henri F., eds., *Existence: A New Dimension Psychiatry and Psychology* (New York: Basic Books, 1960)

Merleau-Ponty, Maurice, *Phenomenology of Perception*, trans. Colin Smith (New York: Humanities Press, 1962)
——, *Sense and Non-Sense*, trans. Hubert L. Dreyfus and Patricia Allen Dreyfus (Evanston: Northwestern University Press, 1964)
——, *Signs*, trans. Richard C. McCleary (Evanston: Northwestern University Press, 1964)
——, *The Primacy of Perception and Other Essays*, James Edie, ed. (Evanston: Northwestern University Press, 1964)
——, *The Structure of Behavior*, trans. Alden L. Fisher (Boston: Beacon Press, 1963)
Minkowski, Eugène, "Approches phénoménologiques (vues par un psychopathologue)", in *L'Évolution psychiatrique*, 27(1962), pp. 433-458
——, "À propos de l'affectivité," in *L'Évolution psychiatrique*, 12(1947), pp. 47-70
——, "Devant une feuille de papier blanche (avant-dernières pensées)," in *Revue Philosophique de Louvain*, 61(1963), pp. 241-281
——, "Esquisses phénoménologiques," in *Recherches philosophiques*, 4(1934), pp. 295-313
——, "Étude psychologique et analyse phénoménologique d'un cas de mélancolie schizophrénique," in *Journal de psychologie normale et pathologique*, 20(1923), pp. 543-560
——, "Expansion et épanouissement," in *Tijdschrift voor Philosophie*, 18(1956), pp. 597-624
——, *La Schizophrénie: Psychopathologie des schizoïdes et des schizophrènes* (Paris: Payot, 1927)
——, *Lived Time: Phenomenological and Psychopathological Studies*, trans. Nancy Metzel (Evanston: Northwestern University Press, 1970)
——, "Petite grammaire phénoménologique (quelques traits de plume)," in *Revue Philosophique de Louvain*, 62(1962), pp. 258-298
——, "Phénoménologie du langage," in *Encyclopédie française* (Larousse), 19(1957), 19, 10-13
——, "Phénoménologie et analyse existentielle en psychopathologie," in *L'Évolution psychiatrique*, 13(1948), pp. 137-185
——, *Traité de psychopathologie* (Paris: Presses Universitaires de France, 1968)
——, *Vers une cosmologie: Fragments philosophiques* (Paris: Aubier, 1936)
Plessner, Helmuth, *Laughing and Crying: A Study of the Limits of Human Behavior*, trans. James Spencer Churchill (Evanston: Northwestern University Press, 1970)
Sartre, Jean-Paul, *Being and Nothingness*, trans Hazel E. Barnes (New York: Philosophical Library, 1956)
——, *Imagination*, trans. Forrest Williams (Ann Arbor: The University of Michigan Press, 1962)
——, *Psychology of the Imagination* (New York: Philosophical Library, 1948)
——, *Sketch for a Theory of the Emotions*, trans. Philip Mairet (London: Methuen and Company, 1962)
Scheler, Max, *On the Nature of Sympathy* (London: Routledge and Kegan Paul, 1954)
Straus, Erwin W., *Phenomenological Psychology: Selected Papers*, trans. Erling Eng (New York: Basic Books, 1966)
Straus, Erwin W., ed., *Phenomenology: Pure and Applied* (Pittsburgh: Duquesne University Press, 1964)
Straus, Erwin W., *The Primary World of Senses: A Vindication of Sensory Experience*, trans. Jacob Needleman (London: The Free Press, 1963)

Index of Names

Index of Subjects